Introduction to
Electrical
Installation
Work

Suitable for the City & Guilds Level 2 2399 Series of qualifications in Environmental Technology Systems and the City & Guilds Level 3 2357 qualification in Installing Electrotechnical Systems.

Introduction to Electrical Installation Work

Third edition

Covers the knowledge units of the
Level 2. City & Guilds Environmental Technology Systems.
Level 3. City & Guilds Diploma in Installing Electrotechnical Systems.

Trevor Linsley

ELSEVIER

AMSTERDAM · BOSTON · HEIDELBERG · LONDON · NEW YORK · OXFORD
PARIS · SAN DIEGO · SAN FRANCISCO · SINGAPORE · SYDNEY · TOKYO

Newnes is an imprint of Elsevier

Newnes

Newnes is an imprint of Elsevier
The Boulevard, Langford Lane, Kidlington, Oxford, OX5 1GB
225 Wyman Street, Waltham, MA 02451, USA

First edition 2007
Second edition 2009
Third edition 2011

British Library Cataloguing-in-Publication Data
A catalogue record for this book is available from the British Library
Library of Congress Cataloging-in-Publication Data
A catalog record for this book is available from the Library of Congress

For information on all Newnes publications
visit our website at www.newnespress.com

ISBN 978-0-08-096940-4

Typeset by MPS Limited, a Macmillan Company, Chennai, India
www.macmillansolutions.com

Printed and bound in Italy
11 12 13 14 15   10 9 8 7 6 5 4 3 2 1

Contents

Preface

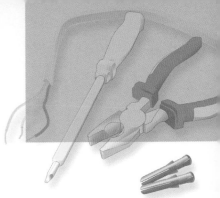

The THIRD edition of *Introduction to Electrical Installation Work* is, as the title implies, a first book of electrical installation practice. It is designed to be a simple introduction to electrical theory and practice and, therefore, does not contain any difficult mathematics or complicated electrical theory.

The book will be of assistance to students taking a first year electrical course, particularly those taking the City and Guilds Level 2 certificate in Environmental Systems and the Level 3 Diploma in Installing Electrotechnical Systems.

Introduction to Electrical Installation Work provides a sound basic knowledge of electrical practice which will also be valuable to the other trades in the construction industry who require a knowledge of electrical installation work, particularly those involved in installing the new environmental technology systems and multi-skilling activities.

This book incorporates the requirements of the latest regulations, particularly:

- The 17th Edition IEE Wiring Regulations (BS 7671:2008)
- New Hazardous Waste Regulations 2005
- The New Work at Height Regulations 2005
- The New Part P Building Regulations (Electrical Safety in Dwellings) 2005
- The New (Harmonised) Fixed Cable Core Colours (2006)

The content of this new edition have been updated and expanded to incorporate the new envirnomental technology subjects in the City and Guilds examinations. Each chapter concludes with Assessment Questions in preparation for the City & Guilds On-Line Assessment.

Margin notes identify:

- KEY FACT which identifies important safety information
- FOLLOW THIS MATHS reminds students to follow each step of a mathematical solution carefully and
- TRY THIS is a student activity which readers may like to respond to by making notes in the margin.

I would like to acknowledge the assistance given by the following manufacturers and organisations in the preparation of this book:

Crabtree Electrical Industries Ltd
R.S. Components Ltd
The Institution of Electrical Engineers
The British Standards Institution
The City & Guilds of London Institute
Stocksigns Ltd
Wylex Electrical Components

I would also like to thank the many college lecturers who returned the questionnaire from Elsevier and the proposal reviewers for their suggestions and advice during the preparation of this book.

I would also like to thank the editorial and production staff at Elsevier for responding to the very short time scale created by the publication of the new City and Guilds qualification.

Finally, I would like to thank Joyce, Samantha and Victoria for their support and encouragement.

Trevor Linsley
2011

CHAPTER **1**

Working effectively and safely in an electrical environment

Online Material WWW

To support your learning, animated versions of some of the figures from this chapter are viewable for free from the book's companion site. When you see the logo, visit the website below to access them: **www.elsevierdirect.com/ companions/9780080969404**

This chapter describes the requirements that are essential to enable electrotechnical activities to be carried out safely and effectively within the parameters set by the current safety legislation and best practice related to the electrotechnical industry.

Laws and safety regulations

The construction industry is one of the biggest industries in the United Kingdom, although most workers are employed by small companies employing less than 25 people.

The construction industry carries out all types of building work from basic housing to offices, hotels, schools and airports.

In all of these construction projects the electrotechnical industry plays a major role in designing and installing the electrical systems to meet the needs of those who will use the completed buildings.

The construction process is potentially hazardous because of the temporary nature of the project and many construction sites these days insist on basic safety standards being met before you are allowed on site. All workers must wear hard hats and safety boots or safety

Introduction to Electrical Installation Work. 978-0-08-096940-4 trainers and use low voltage or battery tools. When the building project is finished, all safety systems will be in place and the building will be safe for those who will use it. However, during the construction period temporary safety systems are in place. People work from scaffold towers, ladders and stepladders. Permanent stairways and safety handrails must be put in by the construction workers themselves.

When the electrical team arrives on site to, let us say, 'first fix' a new domestic dwelling house, the downstairs floorboards and the ceiling plasterboards will probably not be in place, and the person putting in the power cables for the downstairs sockets will need to step over the floor joists, or walk and kneel on planks temporarily laid over the floor joists.

The electrical team spend a lot of time on their hands and knees in confined spaces, on ladders, scaffold towers and on temporary safety systems during the 'first fix' of the process and, as a consequence, slips, trips and falls do occur.

To make all working environments safer, laws and safety regulations have been introduced. To make **your** working environment safe for yourself and those around you, you must obey all the safety regulations that are relevant to your work.

The many laws and regulations controlling the working environment have one common purpose, to make the working environment safe for everyone.

Let us now look at some of these laws and regulations as they apply to the electrotechnical industry.

Statutory laws

Acts of Parliament are made up of statutes. Statutory laws and regulations have been passed by Parliament and have therefore become laws. The City & Guilds syllabus requires that we look at seven statutory regulations.

1. The Health & Safety at Work Act 1974

- The purpose of the HSAWA is to provide the legal framework for stimulating and encouraging high standards of health and safety at work.
- The Act places the responsibility for safety at work on **both** workers and employers.
- The HSAWA is an 'Enabling Act' which allows the Secretary of State to make further regulations and modify existing regulations to create a safe working environment without the need to pass another Act of Parliament.

Key fact

There are seven statutory laws.

2. The Electricity at Work Regulations 1989

- These regulations are made under the Health & Safety at Work Act and are enforced by the Health & Safety Executive (HSE).
- The purpose of the regulations is to 'require precautions to be taken against the risk of death or personal injury from electricity in work activities'.
- An electrical installation wired in accordance with the IEE Regulations BS 7671 will also meet the requirements of the EWR.

3. The Electricity Safety, Quality and Continuity Regulations 2002

- These regulations are designed to ensure a proper and safe supply of electrical energy up to the consumer's mains electrical intake position.
- They will not normally concern the electrical contractor, except in that it is these regulations which set out the earthing requirements of the supply.

4. The Management of Health & Safety at Work Regulations 1999

- To comply with the Health & Safety at Work Act employers must have 'robust health and safety systems and procedures in the workplace'.
- Employers must 'systematically examine the workplace, the work activity and the management of safety through a process of risk assessment'.
- Information based upon the risk assessment findings must be communicated to relevant staff.
- So, risk assessment must form a part of any employer's 'robust policy of health and safety'.

5. The Provision and Use of Work Equipment Regulations 1998

- These regulations place a general duty of care upon employers to ensure minimum requirements of plant and equipment used in work activities.
- If an employer has purchased good quality plant and equipment, and that plant and equipment is well maintained, there is little else to do.

6. The COSHH Regulations 2002

- The Control of Substances Hazardous to Health Regulations (COSHH) control people's exposure to hazardous substances in the workplace.
- Employers must carry out risk assessments and, where necessary, provide PPE (Personal Protective Equipment) so that employees will not endanger themselves.
- Employees must also receive information and training in the safe storage, disposal and emergency procedures which are to be followed by anyone using hazardous substances.

7. The Personal Protective Equipment at Work Regulations 1992 (PPE)

- PPE is defined as all equipment designed to be worn or held in order to protect against a risk to health and safety.
- This includes most types of protective clothing and equipment such as eye, foot and head protection, safety harnesses, life jackets and high visibility clothing.
- Employers must provide PPE free of charge and employees must make use of it for their protection.
- Figure 1.1 below shows the type of safety signs which might be used to indicate the type of PPE to be worn in particular circumstances for your protection.

Try this

Have you seen these or any other PPE signs at work? Make a list of the PPE signs that you have seen at work and state why they were important in that particular work situation. You might like to make notes in the margin here.

Non-statutory regulations

Statute law is law which has been laid down by Parliament as an Act of Parliament.

Non-statutory regulations and codes of practice interpret the statutory regulations.

Non-statutory **does not** mean non-compulsory. If the non-statutory regulation is relevant to your part of the electrotechnical industry then you **must** comply.

Key fact

Non-statutory laws are still very important.

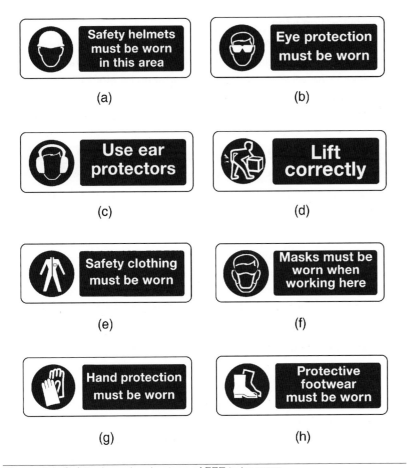

Figure 1.1 Safety signs showing type of PPE to be worn.

The City & Guilds syllabus requires us to look at only one non-statutory regulation, the IEE Regulations.

The IEE Wiring Regulations, the Requirements for Electrical Installations (BS 7671)

- The IEE Wiring Regulations relate principally to the design, selection, erection, inspection and testing of electrical installations. The regulations are non-statutory regulations but are recognised as the national standard for electrical installation work in the United Kingdom.
- **They apply to:**
  - permanent or temporary installations
  - in and about buildings generally
  - agricultural and horticultural premises

- construction sites
- caravans and caravan sites
- They are the 'Electrician's Bible' and provide an authoritative framework for all work activities undertaken by electricians.
- If your work meets the requirements of the IEE Regulations, it will also comply with the statutory regulations.

Health and safety responsibilities

Everyone has a duty of care under the Health and Safety at Work laws and regulations to take care of themselves and others who may be affected by their work activities.

In general terms, the employer must put adequate health and safety systems in place at work and an employee (worker) must use all safety systems and procedures responsibly. In more specific terms, there are **twelve** actions that an employer must take to comply with the health and safety laws.

An employer must:

- make the workplace safe and without risk to health
- provide a Health & Safety Policy Statement if there are more than **five** employees
- provide adequate information, instruction, training and supervision necessary for the health and safety of all employees
- provide any protective clothing or equipment (PPE) required by the Health & Safety at Work Act
- report certain injuries, diseases and dangerous occurrences to the enforcing authorities
- provide adequate first aid facilities
- provide adequate welfare facilities such as toilets and hand washing
- undertake precautions against fire, provide adequate means of escape and the means of fighting a fire
- ensure that plant, equipment and machinery are safe and that safe systems and procedures of work are put in place and followed
- ensure that articles and substances are moved, stored and used safely
- keep dust, fumes and noise under control
- display a current certificate as required by the Employers' Liability (Compulsory Insurance) Act 1969.

An employee (worker) must:

- take reasonable care of their own health and safety and that of others who might be affected by what they do, or may not do, while at work

- co-operate with their employer on all matters relating to health and safety issues
- not interfere with, or misuse anything provided for health and safety or welfare in the workplace
- report any identified health and safety problem in the workplace to a supervisor, manager or employer.

Safety signs

Safety signs are displayed in the working environment to inform workers of the rules and regulations especially relevant to a particular section of the workplace.

They inform and give warning of possible danger and **must be obeyed**.

There are **four types** of safety signs:

1 Warning signs
2 Advisory signs
3 Mandatory signs
4 Prohibition signs.

Warning signs (these give safety information)

These are triangular yellow signs with a black border and symbol as shown in Fig. 1.2.

(a)

(b)

(c)

(d)

Figure 1.2 Warning signs.

Figure 1.3 Advisory or safe condition signs.

Advisory signs (these also give safety information)

Advisory, or safe condition, signs are square or rectangular green signs with a white symbol as shown in Fig. 1.3. They give information about safety provision.

Mandatory signs (these are 'MUST DO' signs)

These are circular blue signs with a white symbol as shown in Fig. 1.4. They give instructions which **must be obeyed**.

Prohibition signs (these are 'MUST NOT DO' signs)

These are circular white signs with a red border and red cross bar as shown in Fig. 1.5. They indicate an activity which **must not** be carried out.

Try this

Have you ever seen any of the signs on these three pages when you have been working? Make a list of where you saw them and why they were important in that work situation. Make some notes in the margin here.

Figure 1.4 Mandatory signs.

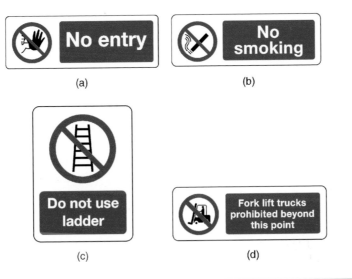

Figure 1.5 Prohibition signs.

Accident and emergency procedures

Despite new legislation, improved information, education and training, accidents at work do still happen.

- An accident may be defined as any uncontrolled event causing injury or damage to an individual or property.
- Make sure that even small accidents at work are recorded in the first aid/accident report book such as that shown in Fig. 1.6.

To avoid having an accident you should:

- recognize situations which could lead to an accident and avoid them
- follow your company's safety procedures – for example, fit safety signs when isolating electricity supplies and screen off work areas from the general public

Figure 1.6 First aid logbook/accident book with data protection compliant removable sheets.

- do not misuse or interfere with equipment provided to protect health and safety
- dress appropriately and use PPE when necessary
- behave appropriately and with care
- stay alert and avoid fatigue
- always work within your level of competence
- take a positive decision to act and work safely. A simple accident may prevent you from working or following your favourite sport or hobby.

Try this

Have you had an accident at work?
Did it hurt? Did you report it? Was it written up in the first aid/accident report book? Was the book data protection compliant with removable sheets?

Emergency procedures – fire control

Fires in industry damage property and materials, injure people and sometimes cause loss of life. Everyone should make an effort to prevent fires, but those which do break out should be extinguished as quickly as possible.

In the event of a fire you should:

- raise the alarm
- turn off machinery, gas and electricity supplies in the area of the fire
- close doors and windows but without locking or bolting them

Key fact

Only attack the fire if you can do so without endangering your own safety in any way.

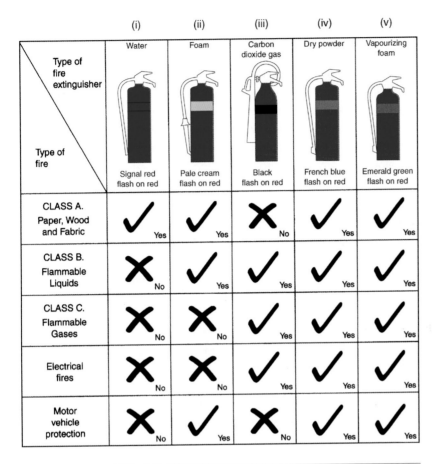

	(i) Water	(ii) Foam	(iii) Carbon dioxide gas	(iv) Dry powder	(v) Vapourizing foam
Type of fire extinguisher / Type of fire	Signal red flash on red	Pale cream flash on red	Black flash on red	French blue flash on red	Emerald green flash on red
CLASS A. Paper, Wood and Fabric	✓ Yes	✓ Yes	✗ No	✓ Yes	✓ Yes
CLASS B. Flammable Liquids	✗ No	✓ Yes	✓ Yes	✓ Yes	✓ Yes
CLASS C. Flammable Gases	✗ No	✗ No	✓ Yes	✓ Yes	✓ Yes
Electrical fires	✗ No	✗ No	✓ Yes	✓ Yes	✓ Yes
Motor vehicle protection	✗ No	✓ Yes	✗ No	✓ Yes	✓ Yes

Figure 1.7 Fire extinguishers and their applications.

- remove combustible material away from the path of the fire if this can be done safely
- attack small fires with the correct extinguisher.

Fires are divided into four classes or categories:

- Class A are wood, paper and textile fires
- Class B are liquid fires such as paint, petrol and oil
- Class C are fires involving gas or spilled liquefied gas
- Class D are very special types of fire involving burning metal.

Electrical fires do not have a special category because, once started, they can be identified as one of the above four types.

Fire extinguishers are for dealing with small fires and different types of fire must be attacked with a different type of extinguisher.

Figure 1.7 shows the correct type of extinguisher to be used on the various categories of fire. The colour coding shown is in accordance with BS EN3:1996.

Emergency procedures – electric shock

- Electric shock occurs when a person becomes part of the electrical circuit as shown in Fig. 1.8.
- The level or intensity of the shock will depend upon many factors such as age, fitness and the circumstances in which the shock is received.
- The lethal level is approximately 50 mA, above which muscles contract, the heart flutters and breathing becomes difficult.
- Below 50 mA only an unpleasant tingling sensation may be experienced or you may feel like you have been struck very hard in the chest.

Actions to be taken upon finding a workmate receiving an electric shock are as follows:

- Switch off the supply if possible
- Alternatively, remove person from the supply *without touching him*
- If breathing or heart has stopped, immediately call professional help by dialling the emergency number 999 (or 112) and asking for the ambulance service. Give precise directions to the scene of the accident. The casualty stands the best chance of survival if the emergency services can get a rapid response paramedic team quickly to the scene. They have extensive training and will have specialist equipment with them
- Only then should you apply resuscitation or cardiac massage until the patient recovers or help arrives.

Figure 1.8 Touching a live conductor can make a person part of the electrical circuit and may lead to an electric shock.

Emergency procedures – first aid

Despite all the health and safety laws and regulations and despite all the safety precautions taken in the working environment to prevent injury to the workforce, accidents do happen at work and if a workmate is injured you will want to help. If you are not a qualified first aider, limit your actions to the obvious commonsense assistance and get help from someone who is qualified.

Let us now look at some first aid procedures which should be practised under expert guidance before they are required in an emergency.

Bleeding

If the wound is dirty, rinse it under clean running water. Clean the skin around the wound and apply a plaster, pulling the skin together.

If the bleeding is severe, apply direct pressure to reduce the bleeding. If the injury is to a limb then raise the limb if possible. Apply a sterile dressing or pad and bandage firmly before obtaining professional advice.

To avoid possible contact with hepatitis or the AIDS virus, when dealing with open wounds, first aiders should avoid contact with fresh

Key fact

To prevent people receiving an electric shock accidentally, all circuits must contain protective devices and all exposed metal must be earthed.

Key fact

All circuits must be electrically isolated before any work is carried out.

blood by wearing plastic or rubber protective gloves, or by allowing the casualty to apply pressure to the bleeding wound.

Burns

Remove heat from the burn to relieve the pain by placing the injured part under clean cold water if at all possible. Do not remove burnt clothing sticking to the skin. Do not apply lotions or ointments. Do not break blisters or attempt to remove loose skin. Cover the injured area with a clean dry dressing.

Broken bones

Make the casualty as comfortable as possible by supporting a broken limb, either by hand or with padding. **Do not move** the casualty unless by remaining in that position they are likely to suffer further injury. Obtain professional help as soon as possible.

Contact with chemicals

Wash the affected area very thoroughly with clean cold water. Remove any contaminated clothing. Cover the affected area with a clean sterile dressing and seek expert advice. It is a wise precaution to treat all chemical substances as possibly harmful; even commonly used substances can be dangerous if contamination is from concentrated solutions. When handling dangerous substances it is also good practice to have a neutralising agent to hand.

Exposure to toxic fumes

Get the casualty into fresh air quickly and encourage deep breathing if conscious. Resuscitate if breathing has stopped. Obtain expert medical advice as fumes may cause irritation of the lungs.

Sprains and bruising

A cold compress can help to relieve swelling and pain. Soak a towel or cloth in cold water, squeeze it out and place it on the injured part. Renew the compress every few minutes.

Breathing stopped

Remove any restrictions from the face and any vomit or loose or false teeth from the mouth. Loosen tight clothing around the neck, chest and waist. To ensure a good airway, lay the casualty on his back and support the shoulders on some padding. Tilt the head backwards and open the mouth. If the casualty is faintly breathing, lifting the tongue clear of the airway may be all that is necessary to restore normal breathing. However, if the casualty does not begin to breathe, open your mouth wide and take a deep breath, close the casualty's nose by pinching with your fingers and, sealing your lips around his mouth, blow into his lungs until the chest rises. Remove your mouth and watch the casualty's chest fall. Continue this procedure at your natural breathing rate. If the mouth is damaged or you have difficulty making a seal around the casualty's mouth, close his mouth and inflate the lungs through the nostrils. Give artificial respiration until natural breathing is restored or until professional help arrives.

Heart stopped beating

This sometimes happens following a severe electric shock. If the casualty's lips are blue, the pupils of his eyes widely dilated and the pulse in the neck cannot be felt, then he may have gone into cardiac arrest. Act quickly and lay the casualty on his back. Kneel down beside him and place the heel of your hand in the centre of his chest. Cover this hand with your other hand and interlace the fingers. Straighten your arms and press down on his chest sharply with the heel of your hands and then release the pressure. Continue to do this **15** times at the rate of **one push per second**. Check the casualty's pulse. If none is felt, give **two breaths** of artificial respiration and then a further **15** chest compressions. Continue this procedure until the heartbeat is restored and the artificial respiration until normal breathing returns. Pay close attention to the condition of the casualty while giving the heart massage. When a pulse is restored the blueness around the mouth will quickly go away and you should stop the heart massage. Look carefully at the rate of breathing. When this is also normal, stop giving artificial respiration. Treat the casualty for shock, place him in the recovery position and obtain professional help.

Shock

Everyone suffers from shock following an accident. The severity of the shock depends upon the nature and extent of the injury. In cases of severe shock the casualty will become pale and his skin will become clammy from sweating. He may feel faint, have blurred vision, feel sick and complain of thirst. Reassure the casualty that everything that needs to be done is being done. Loosen tight clothing and keep the casualty warm and dry until help arrives. **Do not** move the casualty unnecessarily or give anything to drink.

Finally, remember that every accident must be reported to an employer and the details of the accident and treatment given suitably documented. A first aid logbook or accident report book such as that shown in Fig. 1.6 above containing first aid treatment record sheets could be used to effectively document such accidents that occur in the workplace and the treatment given. Failure to do so may influence the payment of compensation at a later date if an injury leads to permanent disability. To comply with the Data Protection Act, from 31 December 2003 all first aid treatment logbooks or accident report books must contain perforated sheets which can be removed after completion and filed away for personal security.

Emergency procedures – electrical isolation and lock off

- The IEE Regulations tell us that every circuit must be provided with a means of isolation.
- The Electricity at Work Regulations tell us that before work commences on electrical equipment it must be disconnected from the source of supply and that the disconnection must be secure.
- A small padlock will ensure the security of the disconnection, or the fuse or MCB may be removed and kept in a safe place whilst work is carried out.
- Where a test instrument or voltage indicator is used to prove the supply dead, the same device must be tested to prove it is still working. Figure 1.9 shows a typical voltage indicator and Fig. 1.10 shows a typical voltage proving unit.

> **Key fact**
>
> Electrical isolation is an important safety procedure – practise this technique at work under the guidance of your supervisor.

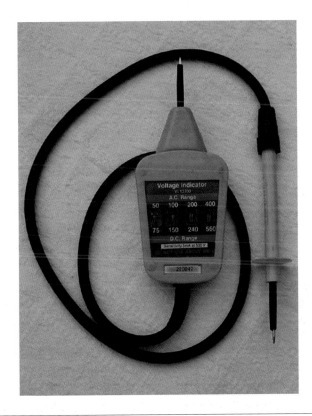

Figure 1.9 Typical voltage indicator.

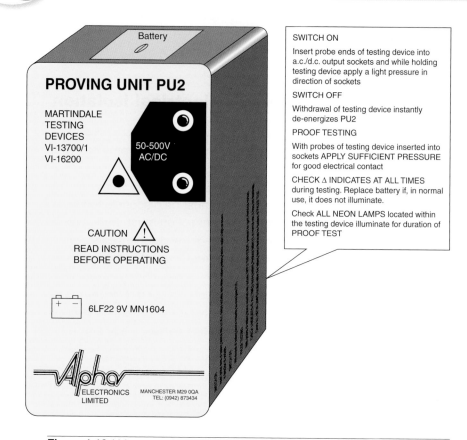

Figure 1.10 Voltage proving unit.

- The test leads and probes of the test instrument must comply with the Health & Safety Executive Guidance Note 38, giving adequate protection to the user. These are robust leads with finger shields and shown in Fig. 1.9.
- To deter anyone from reconnecting the supply, a notice must be fixed on the isolator saying 'Danger – Electrician at Work'.

A suitable electrical isolation procedure is shown in Fig. 1.11, which you should practise in the workshop under the guidance of your lecturer or at work under the guidance of your supervisor. Electrical isolation is an important safety procedure.

Organizations having electrotechnical activities

When we talk about the electrotechnical industry we are referring to all those different organizations or companies which provide an electrical service of some kind.

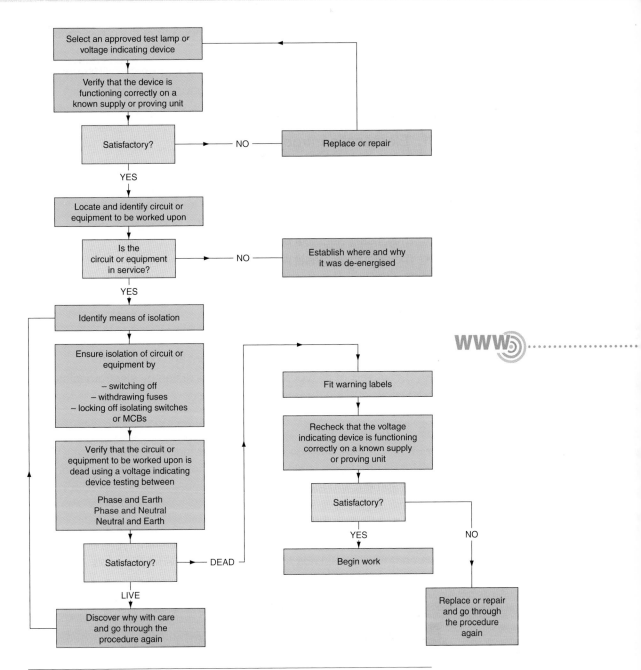

Figure 1.11 Flowchart for a secure isolation procedure.

Electrical contractors install equipment and systems in new buildings. Once a building is fully operational the electrical contractor may provide a maintenance service to the client or customer or alternatively the client may employ an 'in-house' electrician to maintain the installed electrical equipment. It all depends on the amount of work to be done and the complexity of the customer's systems.

The City & Guilds syllabus directs us to look at **twelve** different organizations having electrotechnical activities and ten services provided by the electrotechnical industry, so here goes.

1. Electrical contractors

- Electrical contractors provide a design and installation service for all types of buildings and construction projects
- The focus of this type of organization is on all types of electrotechnical activities in and around buildings
- They install electrical equipment
- They install electrical wiring systems
- They carry out their installation work in domestic, commercial, industrial, agricultural and horticultural buildings.

2. Factories

- Factories contain lots of electrical plant and equipment
- The wheels of all types of industry are driven by electromechanical devices and electrotechnical activities.

3. Process plants

- Whether they process food or nuclear fuels, the prime mover for all processes is electrical plant, control and instrumentation equipment and machine drives.

4. Local councils

- Local councils are responsible for many different types of community buildings from town halls to swimming pools
- The buildings all have electrical systems which require installation, maintenance and repair.

Try this

Which type of electrotechnical organization do you belong to? What types of work have you carried out so far? Make some notes in the margin here.

5. Commercial buildings and complexes

- The 'office type' activities carried out in these buildings require that electrical communication and data transmission systems are installed, maintained and repaired.

The electrotechnical industry is made up of many different organizations or companies.

6. Leisure centres

- These type of buildings contain lots of equipment driven by human sweat but which is also controlled and monitored by electrical and electronic systems
- Leisure centres might contain a swimming pool or 'hot-air' sauna. Both types of electrical installation are considered 'Special Installations' by the IEE Regulations BS:7671.

7. Panel builders

- Panel builders build specialist control, protection and isolation main switchgear systems for commerce and industry

- The panel incorporates the isolation and protection systems required by the electrical installation.

8. Motor rewind and repair

- Electrical motors and their drives usually form an integral part of the industrial system or process
- Electrical motors and transformers sometimes break down or burn out
- An exact new replacement can often be quickly installed
- Alternatively, the existing motor can be rewound and reconditioned by a specialist company if time permits.

9. Railways

- The prime mover for a modern inter-city type electric train is an electric motor
- Electric trains require an infrastructure of electrical transmission lines throughout the network
- All rail movements require signal and control systems
- Railway station buildings contain electrical and electronic installations.

10. The armed forces

- The armed forces operate in harsh, hostile and unpredictable environments
- They need to adapt, modify and repair electrical and electronic systems in a war situation away from their home base and a comfortable well-equipped workshop
- A modern warship can contain as many people as an English village. They need electrotechnical systems to support them and to keep them safe twenty-four hours per day, seven days per week.

11. Hospitals

- Hospitals contain a great deal of high technology equipment
- This equipment requires power and electronic systems which require installation, maintenance and repair
- Life monitoring equipment must continue to operate in a power failure
- Standby electrical supplies are, therefore, often an important part of a hospital's electrical installations.

12. Equipment and machine manufacturers

- White goods, brown goods, computer hardware, motors and transformers are manufactured to meet the increasing demands of the domestic, commercial and industrial markets

- They are manufactured to very high standards and often contain very sophisticated electrical and electronic circuits and systems
- They are manufactured to British and European standards.

Services provided by the electrotechnical industry

1. Lighting and power installations

- Ensure that the building in which they are installed:
  - is illuminated to an appropriate level
  - is heated to a comfortable level
  - has the power circuits to drive the electrical and electronic equipment required by those who will use the buildings.

2. Emergency lighting and security systems

- These ensure that the building is safe to use in unforeseen or adverse situations
- And is secure from unwanted intruders.

3. Building management and control systems

- These systems provide a controlled environment for the people who use commercial buildings
- They provide a pleasant environment so that people can work effectively and efficiently.

Try this

Which services does the company you work for provide within the electrotechnical industry? Make some notes in the margin here.

4. Instrumentation

- Electrical instrumentation allows us to monitor industrial processes and systems often at a safe distance.

5. Electrical maintenance

- A programme of planned maintenance allows us to maintain the efficiency of all installed systems.

6. Live cable jointing

- Making connections to 'live' cables provides a means of connecting new installations and services to existing live supply cables without inconvenience to existing supplies caused by electrical shutdown. **This work requires special training**.

7. Highway electrical systems

- Illuminated motorways, roads and traffic control systems make our roads and pavements safe for vehicles and pedestrians.

8. Electrical panel building

- Main electrical panels provide a means of electrical isolation and protection
- They also provide a means of monitoring and measuring electrical systems in our commercial and industrial buildings.

9. Electrical machine drive installations

- Electrical machine drives drive everything that makes our modern life comfortable from
  - trains and trams to
  - lifts and air conditioning units
  - refrigerators, freezers and all types of domestic appliances.

10. Consumer and commercial electronics

- These give us data processing and number crunching
- Electronic mail and access to information on the world wide web
- Access to high quality audio and video systems.

Roles and responsibilities of workers in the electrotechnical industry

Any electrotechnical organization is made up of a group of individuals with various duties, all working together for their own good, the good of their employer and their customers.

There is often no clear distinction between the duties of the individual employees, each do some of the others' work activities.

Responsibilities vary, even amongst people holding the same job title and some individuals hold more than one job title. However, let us look at some of the roles and responsibilities of those working in the electrotechnical industry.

Design engineer

- Will normally meet with clients and other trade professionals to interpret the customer's requirements
- He or she will produce the design specification which enables the cost of the project to be estimated.

Estimator/cost engineer

- Measures the quantities of labour and material necessary to complete the electrical project using the plans and specifications for the project
- From these calculations and the company's fixed costs, a project cost can be agreed.

Try this

Where do you fit into your company's workforce?
What is your job title – apprentice/trainee?
What is your supervisor's job title?
What is the name of the foreman?

Contracts manager

- May oversee a number of electrical contracts on different sites
- Will monitor progress in consultation with the project manager on behalf of the electrical companies
- Will cost out variations to the initial contract
- May have health & safety responsibilities because he or she has an overview of all company employees and contracts in progress.

Project manager

- Is responsible for the day to day management of one specific contract
- Will have overall responsibility on that site for the whole electrical installation
- Attends site meetings with other trades as the representative of the electrical contractor.

Service manager

- Monitors the quality of the service delivered under the terms of the contract
- Checks that the contract targets are being met
- Checks that the customer is satisfied with all aspects of the project
- The service manager's focus is customer specific while the project manager's focus is job specific.

Technician

- Will be more office-based than site-based
- Will carry out surveys of electrical systems
- Updates electrical drawings
- Obtains quotations from suppliers
- Maintains records such as ISO 9000 quality systems
- Carries out testing, inspections and commissioning of electrical installations
- Troubleshoots.

Supervisor/foreman

- Will probably be a mature electrician
- Has responsibility for small contracts
- Has responsibility for a small part of a large contract
- Will be the leader of a small team (e.g. electrician and trainee) installing electrical systems.

Operative

- Carries out the electrical work under the direction and guidance of a supervisor
- Should demonstrate a high degree of skill and competence in electrical work
- Will have, or be working towards, a recognized electrical qualification and status as an electrician, approved electrician or electrical technician.

Mechanic/fitter

- An operative who usually has a 'core skill' or 'basic skill' and qualification in mechanical rather than electrical engineering
- In production or process work he or she would have responsibility for the engineering and fitting aspects of the contract, while the

electrician and instrumentation technician would take care of the electrical and instrumentation aspects

- All three operatives must work closely in production and process work
- 'Additional skilling' or 'multi-skilling' training produces a more flexible operative for production and process plant operations.

Maintenance manager/engineer

- Is responsible for keeping the installed electrotechnical plant and equipment working efficiently
- Takes over from the builders and contractors the responsibility of maintaining all plant, equipment and systems under his or her control
- Might be responsible for a hospital or a commercial building, a university or college complex
- Will set up routine and preventative maintenance programmes to reduce possible future breakdowns
- When faults or breakdowns do occur he or she will be responsible for the repair using the company's maintenance staff.

People definitions

The IEE Regulations describe peoples' skills and abilities to work safely in the following way.

A competent person is one who has sufficient technical knowledge, relevant practical skills and experience to be able to perform a particular task properly and safely. Generally speaking an electrician will have the necessary skills to perform a wide range of electrical activities competently.

The HSE Regulation 16 states that persons 'must be competent to prevent danger ..... so that the person themselves or others are not placed at risk due to a lack of skill when dealing with electrical equipment.'

A skilled person is a person with technical knowledge or sufficient experience to be able to avoid the dangers which electricity may create. NVQ level 3 is 'skilled craft level' or the level to be considered 'competent'.

An instructed person is a person adequately advised or supervised by skilled persons to be able to avoid the dangers which electricity may create.

An ordinary person is a person who is neither a skilled person nor an instructed person.

Try this

A person may be described as:

- *ordinary*
- *competent*
- *instructed or*
- *skilled*

depending upon that person's skill or ability.
Put people you know into each category, for example yourself, your supervisor, your parents, your friends. Make notes in the margin here to help with this activity.

Professional bodies supporting electrotechnical organizations

If you are reading this book I would guess that you are an electrical trainee working in one sector of the electrotechnical industry. You hope to eventually pass the City & Guilds 2357 NVQ3 qualifications, take your AM2 Practical Assessment and become a qualified electrician. Believe me, I do wish you well, because you are the future of the electrotechnical industry.

As a trainee, you are probably employed by an electrical company and attend your local college on either a 'Day Release' or 'Block Release' scheme. The combination of work and college will provide you with the skills you will need to become 'fully qualified'!

So, although you are doing all the work yourself, you are being sponsored or supported by the company that you work for, the JTL (JIB Training Limited) and the City & Guilds of London Institute to become professionally qualified as an electrician.

It is in this same way that the professional bodies support the electrotechnical industry. They provide a structure of help, support and guidance to the individual companies that make up the electrotechnical industry.

So let us look at some of the professional bodies which support the electrotechnical organizations like the company you work for.

The IET (The Institution of Engineering and Technology)

- The IET was formed in spring 2006 by bringing together the IEE (Institution of Electrical Engineers) and the IIE (Institution of Incorporated Engineers)

- The IET is Europe's largest professional society for engineers
- The IET publishes the IEE Wiring Regulations to BS: 7671
- It also produces many other publications and provides training courses to help electricians, managers and supervisors to keep up to date with the changes in the relevant regulations
- The IEE On Site Guide describes the 'requirements for electrical installations'
- Eight guidance notebooks are available
- The Electrician's Guide to the Building Regulations clarifies the requirements for electrical operatives of the new Part P Regulations which came into effect on 1 January 2005
- *Wiring Matters* is a quarterly magazine published by the IET covering many of the topics which may trouble some of us in the electrotechnical industry
- All of these publications can be purchased by visiting the IET website at www.theiet.org.uk/shop.

The ECA (Electrical Contractors Association)

- The ECA was founded over 100 years ago and is a trade association representing electrotechnical companies
- Membership is made up of electrical contracting companies both large and small
- Customers employing an electrical contractor who has ECA membership are guaranteed that the work undertaken will meet all relevant regulations. If the work undertaken fails to meet the relevant standards, the ECA will arrange for the work to be rectified at no cost to the customer
- The work of the ECA member is regularly assessed by the Association's UKAS accredited inspection body
- Those electrotechnical companies which are members of the ECA are permitted to display the ECA logo on their company vehicles and stationery
- Further information can be found on the ECA website at www.eca.co.uk.

The National Inspection Council for Electrical Installation Contracting (NICEIC)

- The NICEIC is an independent consumer safety organization, set up to protect users of electricity against the hazards of unsafe electrical installations

- It is the electrical industry's safety regulatory body
- The NICEIC publishes a list of approved contractors whose standard of work is regularly assessed by local area engineers
- Customers employing an electrical contractor who has NICEIC membership can be assured that the work carried out will meet all relevant standards. If the work undertaken fails to meet all relevant standards, the name of the electrical contractor will be removed from the 'NICEIC Approved List'
- Some work, such as local authority work, is only available to NICEIC approved contractors
- Further information can be found at www.niceic.org.uk.

Trade unions

- Trade unions have a long history of representing workers in industry and commerce
- The relevant unions negotiate with employer organizations the pay and working conditions of their members
- The trade union which represents employees in the electrotechnical industry in the new millennium is called Unite
- Through a network of local area offices the union offers advice and support for its members. They will also provide legal advice and representation if a member has a serious accident as a result of a health and safety issue or has a dispute with an employer
- Further information can be found at www.unitetheunion.org.

! Try this

Does the company you work for belong to a trade organization?
Why do they belong, what are the advantages?
Do you belong to a trade union – if so, which one?
If not, why? Trade union membership is often free while you are training.

Communications

When we talk about good communications we are talking about transferring information from one person to another both quickly and accurately. We do this by talking to other people, looking at drawings and plans and discussing these with colleagues from the same company and with other professionals who have an interest in the same project. The technical information used within our industry comes from many sources. The IEE Regulations (BS 7671) are the 'electrician's bible' and form the basis of all our electrical design calculations and installation methods. British Standards, European Harmonized Standards and Codes of Practice provide detailed information for every sector of the electrotechnical industry, influencing all design and build considerations.

Sources of technical information

Equipment and accessories available to use in a specific situation can often be found in the very comprehensive manufacturers' catalogues and the catalogues of the major wholesalers that service the electrotechnical industry.

All of this technical information may be distributed and retrieved by using:

- conventional drawings and diagrams which we will look at in more detail below
- sketch drawings to illustrate an idea or the shape of say a bracket to hold a piece of electrical equipment
- the Internet can be used to download British Standards and Codes of Practice
- the Internet can also be used to download health and safety information from the Health & Safety Executive at www.gov.uk/hse or www.opsi.gov.uk
- CDs, DVDs, USB memory sticks and email can be used to communicate and store information electronically
- the facsimile (fax) machine can be used to communicate with other busy professionals, information say about a project you are working on together

If you are working at your company office with access to online computers, then technical information is only a fingertip or mouse click away. However, a construction site is a hostile environment for a laptop and so a hard copy of any data is preferable on site.

Let us now look at the types of drawings and diagrams which we use within our industry to communicate technical information between colleagues and other professionals. The type of diagram to be used in any particular situation is the one which most clearly communicates the desired information.

Site plans or layout drawings

These are scale drawings based upon the architect's site plan of the building and show the position of the electrical equipment which is to be installed. The electrical equipment is identified by a graphical

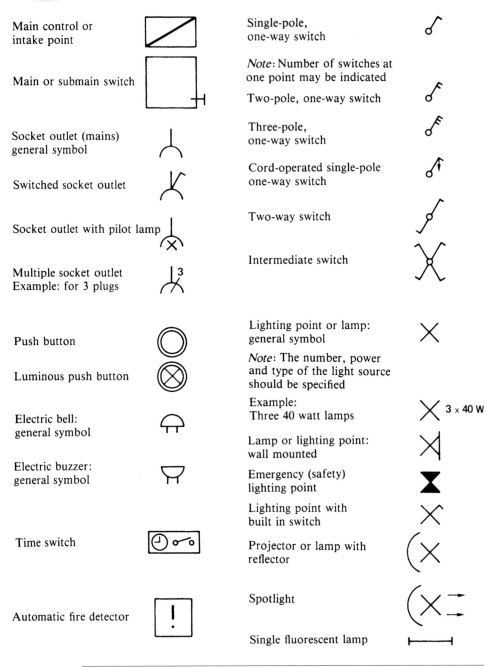

Main control or intake point

Main or submain switch

Socket outlet (mains) general symbol

Switched socket outlet

Socket outlet with pilot lamp

Multiple socket outlet
Example: for 3 plugs

Push button

Luminous push button

Electric bell: general symbol

Electric buzzer: general symbol

Time switch

Automatic fire detector

Single-pole, one-way switch

Note: Number of switches at one point may be indicated

Two-pole, one-way switch

Three-pole, one-way switch

Cord-operated single-pole one-way switch

Two-way switch

Intermediate switch

Lighting point or lamp: general symbol

Note: The number, power and type of the light source should be specified

Example:
Three 40 watt lamps

Lamp or lighting point: wall mounted

Emergency (safety) lighting point

Lighting point with built in switch

Projector or lamp with reflector

Spotlight

Single fluorescent lamp

Figure 1.12 Some BS EN 60617 electrical installation symbols.

symbol. The standard symbols used by the electrical contracting industry are those recommended by the British Standard EN 60617, *Graphical Symbols for Electrical Power, Telecommunications and Electronic Diagrams*. Some of the more common electrical installation symbols are given in Fig. 1.12.

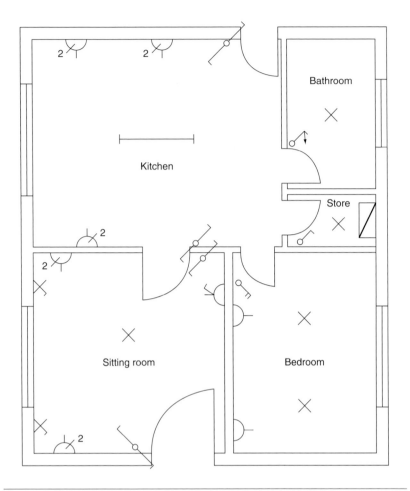

Figure 1.13 Layout drawing or site plan of a small electrical installation.

Try this

The next time you are on site, ask your supervisor to show you the site plans. Ask him to:

- *show you how the scale works*
- *put names to the equipment represented by British Standard symbols.*

The site plan or layout drawing will be drawn to a scale, smaller than the actual size of the building, so to find the actual measurement, you must measure the distance on the drawing and multiply by the scale.

For example, if the site plan is drawn to a scale of 1:100, then 10 mm on the site plan represents 1 m measured in the building.

The layout drawing or site plan of a small domestic extension is shown in Fig. 1.13. It can be seen that the mains intake position, probably a consumer unit, is situated in the storeroom which also contains one light controlled by a switch at the door. The bathroom contains one lighting point controlled by a one-way pull switch at the door. The kitchen has two doors and a switch is installed at each door to control the fluorescent luminaire. There are also three double sockets situated around the kitchen. The sitting room has a two-way switch at each door controlling the centre lighting point. Two wall lights with built-in switches are to be wired, one at each side of the window. Two double sockets and one switched socket are also to be installed in the sitting room. The bedroom has two lighting points controlled independently by two one-way switches at the door. The wiring diagrams and installation procedures for all these circuits can be found in later chapters.

As-fitted drawings

When the installation is completed, a set of drawings should be produced which indicate the final positions of all the electrical equipment. As the building and electrical installation progresses, it is sometimes necessary to modify the positions of equipment indicated on the layout drawing because, for example, the position of a doorway has been changed. The layout drawings or site plans indicate the original intentions for the position of equipment, while the 'as-fitted' drawing indicates the actual positions of equipment upon completion of the contract.

Detail drawings and assembly drawings

These are additional drawings produced by the architect to clarify some point of detail. For example, a drawing might be produced to give a fuller description of a suspended ceiling arrangement or the assembly arrangements of the metalwork for the suspended ceiling.

Try this

Take a moment to clarify the difference between:

- *layout drawings and*
- *as-fitted drawings.*

Make notes in the margin here or highlight the relevant text.

Location drawings

Location drawings identify the place where something is located. It might be the position of the manhole covers giving access to

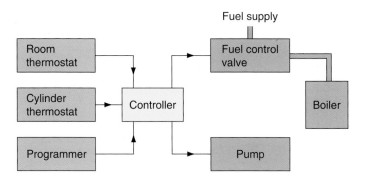

Figure 1.14 Block diagram – space heating control system (Honeywell Y. Plan).

the drains. It might be the position of all water stop taps or the position of the emergency lighting fittings. This type of information may be placed on a blank copy of the architect's site plan or on a supplementary drawing.

Distribution cable route plans

On large installations there may be more than one position for the electrical supplies. Distribution cables may radiate from the site of the electrical mains intake position to other sub-mains positions. The site of the sub-mains and the route taken by the distribution cables may be shown on a blank copy of the architect's site plan or on the electrician's 'as-fitted' drawings.

Block diagrams

A block diagram is a very simple diagram in which the various items or pieces of equipment are represented by a square or rectangular box. The purpose of the block diagram is to show how the components of the circuit relate to each other and, therefore, the individual circuit connections are not shown. Figure 1.14 shows the block diagram of a space heating control system.

Wiring diagrams

A wiring diagram, or connection diagram, shows the detailed connections between components or items of equipment. It does not indicate how a piece of equipment or circuit works. The purpose of a wiring diagram is to help someone with the actual wiring of the circuit. Figure 1.15 shows the wiring diagram for a space heating control system. Other wiring diagrams can be seen in Figs 4.8 and 4.9 (see Chapter 4).

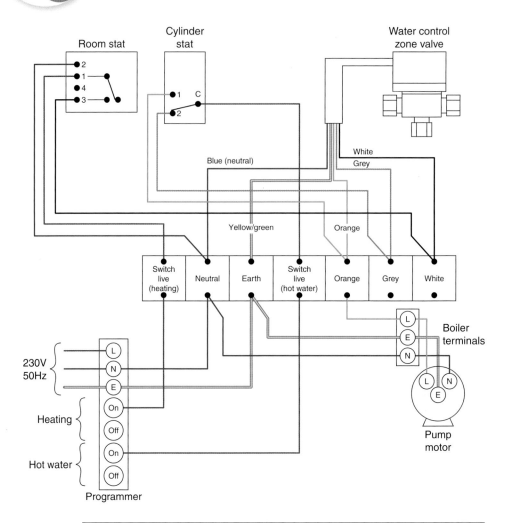

Figure 1.15 Wiring diagram – space heating control system (Honeywell Y. Plan).

Circuit diagrams

A circuit diagram shows most clearly how a circuit works. All the essential parts and connections are represented by their graphical symbols. The purpose of a circuit diagram is to help our understanding of the circuit. It will be laid out as clearly as possible, without regard to the physical layout of the actual components and, therefore, it may not indicate the most convenient way to wire the circuit. Figure 1.16 shows the circuit diagram of our same space heating control system. Figs 2.4 and 2.5 in Chapter 2 are circuit diagrams.

Schematic diagrams

A schematic diagram is a diagram in outline of, for example, a motor starter circuit. It uses graphical symbols to indicate the interrelationship

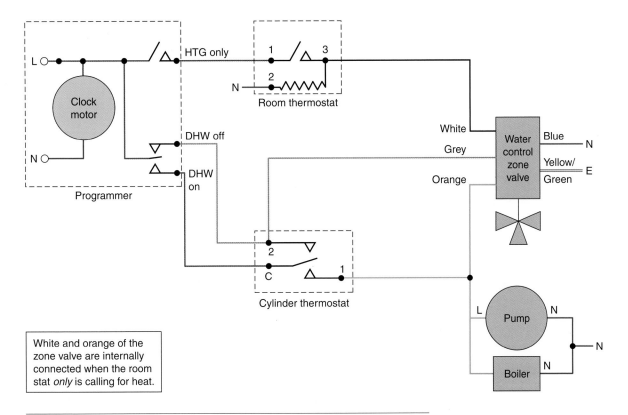

Figure 1.16 Circuit diagram – space heating control system (Honeywell Y. Plan).

of the electrical elements in a circuit. These help us to understand the working operation of the circuit but are not helpful in showing us how to wire the components. An electrical schematic diagram looks very like a circuit diagram. Figure 1.17 shows a schematic diagram.

Freehand working diagrams

Freehand working drawings or sketches are another important way in which we communicate our ideas. The drawings of the spring toggle bolt in Chapter 4 (Fig. 4.33) were done from freehand sketches. A freehand sketch may be done as an initial draft of an idea before a full working drawing is made. It is often much easier to produce a sketch of your ideas or intentions than to describe them or produce a list of instructions.

To convey the message or information clearly, it is better to make your sketch large rather than too small. It should also contain all the dimensions necessary to indicate clearly the size of the finished object depicted by the sketch.

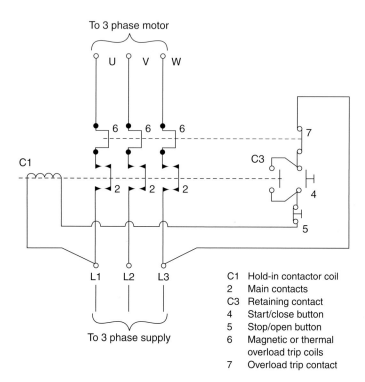

To 3 phase motor

C1   C3

C1	Hold-in contactor coil
2	Main contacts
C3	Retaining contact
4	Start/close button
5	Stop/open button
6	Magnetic or thermal overload trip coils
7	Overload trip contact

Figure 1.17 Schematic diagram – DOL motor starter.

The positional reference system

A positional reference system can be used to mark exact positions in any space. It uses a simple grid reference system to mark out points in the space enclosed by the grid. It is easy to understand if we consider a specific example which I use when building prototype electronic circuits on matrix board. Matrix board is the insulated board full of holes into which we insert small pins and then attach the electronic components.

To set up the grid reference, count along the columns at the top of the board, starting from the left, and then count down the rows. The position of point 4:3 would be 4 holes from the left and 3 holes down.

Prepare a matrix board, or any space for that matter, as follows:

- Turn the matrix board so that a manufactured straight edge is to the top and left-hand side
- Use a felt tip pen to mark the holes in groups of five along the top edge and down the left-hand edge as shown in Fig. 1.18
- The pins can then be inserted as required. Figure 1.18 shows a number of pin reference points. Counting from the left-hand side of the board there are 3:3, 3:16, 10:11, 18:3, 18:11, 25:3 and 25:16.

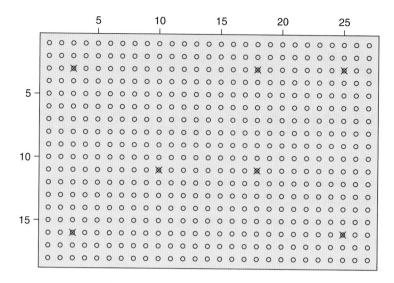

Figure 1.18 Positional reference system used to identify points on an electronic matrix board.

Assessment questions

Identify the statements as true or false. If only part of the statement is false, tick false.

1   The Health & Safety at Work Act provides the legal framework for encouraging good standards of health and safety at work.
    True ☐   False ☐

2   The Wiring Regulations (BS 7671) are recognised as the national standard for electrical work in the UK.
    True ☐   False ☐

3   The Health & Safety at Work Act is a statutory regulation.
    True ☐   False ☐

4   The Wiring Regulations (BS 7671) are statutory regulations.
    True ☐   False ☐

5   An employee – that is a worker – is responsible for his or her own safety at work and the safety of others who might be affected by what they do.
    True ☐   False ☐

6   There are four types of safety signs which inform workers of the rules and regulations especially relevant to a particular work situation.
    Warning signs describe what <u>must not</u> be done.
    True ☐   False ☐

7   Mandatory signs give information which <u>must be obeyed</u> in the work environment.
    True ☐   False ☐

8   Site plans or layout drawings are scaled drawings showing the position of the electrical equipment to be installed.
    True ☐   False ☐

9   Block diagrams show the detailed connections between components or pieces of equipment.
    True ☐   False ☐

10 The ECA is a trade organization representing electrotechnical companies and Unite is a trade union representing employees.
    True ☐   False ☐

Multiple choice assessment questions

Tick the correct answer. Note that more than ONE answer may be correct.

11 Identify the regulations which are statutory regulations

   a.   The HSAWA (Health & Safety at Work Act) ☐

   b.   E @ Work Regs. (Electricity at Work Regulations) ☐

   c.   COSHH (Control of Substances Hazardous to Health) ☐

   d.   IEE Wiring Regs. (BS 7671) ☐

12 Identify the regulations which are non-statutory regulations

   a.   The HSAWA (Health & Safety at Work Act) ☐

   b.   E @ Work Regs. (Electricity at Work Regulations) ☐

   c.   COSHH (Control of Substances Hazardous to Health) ☐

   d.   IEE Wiring Regs. (BS 7671) ☐

13 The HSAWA puts the responsibility for safety at work upon:

   a.   an employee ☐

   b.   an employer ☐

   c.   everyone ☐

   d.   the government ☐

14 To work safely and care for the safety of others is the responsibility of:

   a.   an employee ☐

   b.   an employer ☐

   c.   everyone ☐

   d.   the government ☐

15 To prepare a Health & Safety Policy Statement is the responsibility of:

   a.   an employee ☐

   b.   an employer ☐

c.   everyone ☐

d.   the government ☐

16   Triangular yellow safety signs with a black border and symbol are called:

a.   Advisory signs giving safety information ☐

b.   Mandatory signs or **must do** signs ☐

c.   Prohibition signs or **must not do** signs ☐

d.   Warning signs giving safety information ☐

17   Square or rectangular green signs with a white symbol are called:

a.   Advisory signs giving safety information ☐

b.   Mandatory signs or **must do** signs ☐

c.   Prohibition signs or **must not do** signs ☐

d.   Warning signs giving safety information ☐

18   Circular blue signs with a white symbol are called:

a.   Advisory signs giving safety information ☐

b.   Mandatory signs or **must do** signs ☐

c.   Prohibition signs or **must not do** signs ☐

d.   Warning signs giving safety information ☐

19   Circular white signs with a red border and red cross bar are called:

a.   Advisory signs giving safety information ☐

b.   Mandatory signs or **must do** signs ☐

c.   Prohibition signs or **must not do** signs ☐

d.   Warning signs giving safety information ☐

20   An uncontrolled event causing injury or damage is one definition of:

a.   a runaway bus ☐

b.   a first aid procedure ☐

c.   an accident ☐

d.   an emergency procedure ☐

21   A fire extinguisher showing a signal red flash on a red background contains:

a.   Carbon dioxide gas ☐

b.   Dry powder ☐

c.   Foam ☐

d.   Water ☐

22 **A fire extinguisher showing a black flash on a red background contains:**

a. Carbon dioxide gas ☐

b. Dry powder ☐

c. Foam ☐

d. Water ☐

23 **A fire extinguisher showing a pale cream flash on a red background contains:**

a. Carbon dioxide gas ☐

b. Dry powder ☐

c. Foam ☐

d. Water ☐

24 **Following every accident at work:**

a. an employee must take three days off work ☐

b. a waterproof plaster must be placed on the injury ☐

c. a record must be made in the Accident/First Aid book ☐

d. a report must be sent to the HSE local area office ☐

25 **The Electricity at Work Regulations tell us that before work commences on electrical equipment it must be disconnected from the source of supply and that disconnection must be secure. To comply with this regulation we must:**

a. switch off the circuit at the local functional switch ☐

b. switch off the current at the local isolator switch ☐

c. follow a suitable electrical isolation procedure ☐

d. follow the test procedures given in Part 7 of the IEE Regulations (BS 7671) ☐

26 **Emergency lighting and security systems ensure that a building:**

a. is safe to use in unforeseen circumstances ☐

b. is illuminated and heated to an appropriate level ☐

c. ensures the efficiency of the installed system ☐

d. provides safe monitoring of industrial processes and systems ☐

27 **Electrical maintenance:**

a. is safe to use in unforeseen circumstances ☐

b. is illuminated and heated to an appropriate level ☐

c. ensures the efficiency of the installed system ☐

d. provides safe monitoring of industrial processes and systems ☐

28   The supervisor/foreman will:

a.   oversee a number of electrical contracts   ☐

b.   be responsible for the day to day management of one specific contract   ☐

c.   be the leader of a small team installing electrical systems   ☐

d.   be an operative who has a basic skill and qualification in mechanical rather than electrical engineering   ☐

29   The contracts manager of a company will:

a.   oversee a number of electrical contracts   ☐

b.   be responsible for the day to day management of one specific contract   ☐

c.   be the leader of a small team installing electrical systems   ☐

d.   be an operative who has a basic skill and qualification in mechanical rather than electrical engineering   ☐

30   A mechanic/fitter will:

a.   oversee a number of electrical contracts   ☐

b.   be responsible for the day to day management of one specific contract   ☐

c.   be the leader of a small team installing electrical systems   ☐

d.   be an operative who has a basic skill and qualification in mechanical rather than electrical engineering   ☐

31   A trade union is:

a.   the electrical industry's safety regulatory body   ☐

b.   a professional body supporting electrotechnical organizations   ☐

c.   the British Standard for electrical power supplies   ☐

d.   an organization representing electrical employees   ☐

32   The National Inspection Council for Electrical Installation Contracting is:

a.   the electrical industry's safety regulatory body   ☐

b.   a professional body supporting electrotechnical organizations   ☐

c.   the British Standard for electrical power supplies   ☐

d.   the trade union representing electrical employees   ☐

33   A scale drawing showing the position of equipment by graphical symbol is a description of a:

a.   block diagram   ☐

b.   layout diagram or site plan   ☐

c.   wiring diagram   ☐

d.   circuit diagram   ☐

34 **A diagram which shows the detailed connection between individual items of equipment is a description of a:**

   a.   block diagram   ☐

   b.   layout diagram or site plan   ☐

   c.   wiring diagram   ☐

   d.   circuit diagram   ☐

35 **A diagram which shows most clearly how a circuit works, with all items represented by graphical symbols, is a description of:**

   a.   block diagram   ☐

   b.   layout diagram or site plan   ☐

   c.   wiring diagram   ☐

   d.   circuit diagram   ☐

CHAPTER **2**

Basic principles of electrotechnology

Online Material WWW

To support your learning, animated versions of some of the figures from this chapter are viewable for free from the book's companion site. **When you see the logo, visit the website below to access them:**
www.elsevierdirect.com/ companions/9780080969404

This chapter describes the basic scientific concepts and electrical circuits which form the foundations of electrotechnology.

Basic units used in electrotechnology

In all branches of science, engineering and electrotechnology we use the international metric system of units called the Système International, abbreviated to SI system.

Table 2.1 describes some of the basic units that we shall be using in this chapter.

Like all metric systems, SI units may be increased or reduced by using multiples or sub-multiples of 10. Some of the more common multiples and their names are shown in Table 2.2.

The unit of electrical power is the watt, symbol W, but this is a small unit of power and a more common unit is the kilowatt or one thousand watts. This is expressed as kW in the SI system of units.

Try this

Can you think of other multiples and sub-multiples of basic units? Ask your workmates, work supervisor or college lecturer.

Introduction to Electrical Installation Work. 978-0-08-096940-4

Table 2.1 Basic SI units

Quantity	Measure of	Basic Unit	Symbol	Notes
area	length × length	metre squared	m^2	
current I	electric current	ampere	A	
energy	ability to do work	Joule	J	Joule is a very small unit. 3.6×10^6 J = 1 kWh
force	the effect on a body	Newton	N	
frequency	number of cycles	Hertz	Hz	mains frequency is 50 Hz
length	distance	metre	m	
mass	amount of material	kilogram	kg	1 metric tonne = 1000 kg
magnetic flux Φ	magnetic energy	Weber	Wb	
magnetic flux density B	number of lines of magnetic flux	Tesla	T	
potential or pressure	voltage	volt	V	
period T	time taken to complete one cycle	second	s	the 50 Hz mains supply has a period of 20 ms
power	rate of doing work	Watt	W	
resistance	opposition to current flow	Ohm	Ω	
resistivity	resistance of a sample piece of material	Ohm metre	ρ	resistivity of copper is 17.5×10^{-9} Ωm
temperature	hotness or coldness	Kelvin	K	0°C = 273 K. A change of 1 K is the same as 1°C
time	time	second	s	60 s = 1 min 60 min = 1 h
weight	force exerted by a mass	kilogram	kg	1000 kg = 1 tonne

Note: A more detailed description can be found in Chapter 9 of *Basic Electrical Installation Work* 6th Edition, ISBN 978-0-08-096628-1.

Electrical theory

- All matter is made up of atoms.
- All atoms are made up of a central positively charged nucleus surrounded by negatively charged electrons.
- The electrical properties of materials depend largely upon how tightly these electrons are bound to the nucleus.
- **A conductor** is a material in which the electrons are loosely bound to the central nucleus and, in fact, can very easily become free

Table 2.2 Symbols and multiples for use with SI units

Prefix	Symbol	Multiplication factor		
Mega	M	$\times 10^6$	or	$\times 1000000$
Kilo	k	$\times 10^3$	or	$\times 1000$
Hecto	h	$\times 10^2$	or	$\times 100$
Deca	da	$\times 10$	or	$\times 10$
Deci	d	$\times 10^{-1}$	or	$\div 10$
Centi	c	$\times 10^{-2}$	or	$\div 100$
Milli	m	$\times 10^{-3}$	or	$\div 1000$
Micro	μ	$\times 10^{-6}$	or	$\div 1000000$

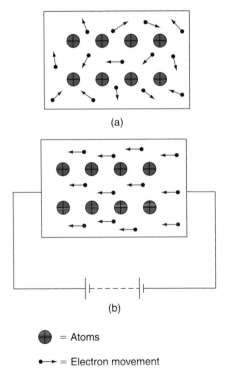

(a)

(b)

⊕ = Atoms

•→ = Electron movement

Figure 2.1 Atoms and electrons in a material. (a) Shows the random movement of free electrons; (b) shows the free electrons drifting towards the positive terminal when a voltage is applied.

electrons. These free electrons drift around randomly inside a conductor as shown in Fig. 2.1(a).

- Good conductors are gold, silver, copper, aluminium, brass etc.
- **An insulator** is a material in which the electrons are very tightly or strongly bound to the central nucleus.
- Good insulators are PVC, rubber, perspex, glass, wood, porcelain etc.

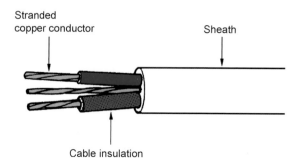

Figure 2.2 A twin and earth PVC insulated and sheathed cable.

Electron flow or electric current

- If a battery is attached to a 'good conductor' material, the free electrons drift toward the positive terminal as shown in Fig. 2.1(b).
- The drift of electrons within a conductor is what we know as an electric current flow.
- Current flow is given the symbol I and is measured in amperes.

Electrical cables

Electrical cables are used to carry electric currents.

Most cables are constructed in three parts:

1 The conductor that carries the current and may have a stranded or solid core.
2 The insulation that contains the current and is colour coded for identification.
3 The outer sheath that may contain some means of providing protection from mechanical damage.

Figure 2.2 shows a PVC insulated and sheathed cable, the type used for domestic installations.

Figure 2.3 shows a PVC/SWA (PVC insulated steel wire armoured) cable, the type used for industrial or underground installations where some mechanical protection is required.

Three effects of an electric current

When an electric current flows in a circuit it can have one or more of the following three effects: **heating**, **magnetic** or **chemical**.

Heating effect

- The electrons moving in the conductor cause the conductor to heat up

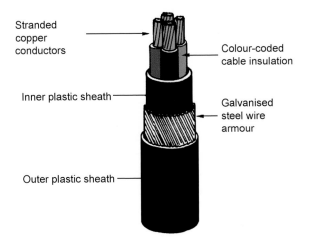

Stranded copper conductors

Colour-coded cable insulation

Inner plastic sheath

Galvanised steel wire armour

Outer plastic sheath

Figure 2.3 A four core PVC/SWA cable.

- The amount of heat generated depends upon the:
  1 amount of current flowing
  2 dimensions of the conductors
  3 type of conductor material used
- Practical applications of the heating effect of an electric current are:
  1 radiant heaters which heat rooms
  2 circuit protection fuses and MCBs which cut off the supply when an overcurrent flows.

Magnetic effect

- Whenever a current flows in a conductor a magnetic field is set up around the conductor like an extension of the insulation – more about this later
- Increasing the current increases the magnetic field
- Switching the current off causes the magnetic field to collapse
- Practical applications of the magnetic effect are:
  1 electric motors which rotate because of the magnetic flux generated by the electrical supply
  2 door chimes and buzzers which ding dong or buzz because of the magnetic flux generated by the electrical supply.

Chemical effect

- When an electric current flows through a conducting liquid, the liquid separates into its chemical parts, a process called electrolysis
- Alternatively, if two metals are placed in a conducting liquid they react chemically and produce a voltage

- Practical applications of the chemical effect are:
  1 industrial processes such as electroplating which is used to silver plate sports trophies and cutlery
  2 motor car batteries which store electrical energy.

Ohm's law

This is one of the most famous electrical laws, published by Dr George Ohm in 1826. It allows us to understand the relationship between the basic elements of an electric circuit—voltage, current and resistance. Voltage is the pressure or potential driving current around a circuit. Current, as we saw a little earlier at Fig. 2.1, is the movement of electrons through a conductor and resistance is the opposition to that current flow. Ohm's law may be expressed as *voltage is equal to current times resistance* or expressed mathematically as:

$$V = I \times R \text{ volts}$$

Transposing this formula, we have:

$$\text{Current } I = \frac{V}{R}\text{(A)} \quad \text{and} \quad \text{Resistance } R = \frac{V}{I}\text{(Ω)}$$

Follow this maths

Follow this maths carefully step by step.

Example 1

An electric fan heater was found to take **10A** when connected to the **230V** mains supply. Calculate the resistance of the heater element.

$$\text{From } R = \frac{V}{I}\text{(Ω)}$$

$$R = \frac{230V}{10A} = 23\text{(Ω)}$$

The heater element resistance is 23 ohm

Example 2

Calculate the current flowing in a disco 'sound and light' unit having a resistance of **57.5 Ω** when it is connected to the **230 V** electrical mains.

$$\text{From } I = \frac{V}{R}\text{(A)}$$

$$I = \frac{230V}{57.5\Omega} = 4\text{(A)}$$

The 'sound and light' unit takes 4 amps

Table 2.3 Resistivity values

Material	Resistivity (Ωm)
Silver	16.4×10^{-9}
Copper	17.5×10^{-9}
Aluminium	28.5×10^{-9}
Brass	75.0×10^{-9}
Iron	100.0×10^{-9}

Resistivity

The resistance or opposition to current flow varies, depending upon the type of material being used to carry the electric current.

Resistivity is defined as the resistance of a sample of a particular material and Table 2.3 gives the resistivity values of some common materials.

Using these values we can calculate the resistance of different materials using the formula:

$$\text{Resistance } R = \frac{\rho l}{a} (\Omega)$$

where ρ (the Greek letter rho) is the resistivity value for the material, l is the length and a is the cross-sectional area.

Example 3

Calculate the resistance of **100 m** of **2.5 mm^2** copper cable using the resistivity values in Table 2.3.

$$\text{We know that } R = \frac{\rho l}{a} (\Omega)$$

$$\text{Therefore } R = \frac{17.5 \times 10^{-9} \times 100}{2.5 \times 10^{-6}}$$

$$\therefore R = 700 \times 10^{-3} (\Omega) \text{ or}$$
$$R = 700 \text{ (m}\Omega)$$

Note: the cross section of the cable is in mm^2;
mm $= 10^{-3}$ (see Table 2.2) so, mm$\times$mm $= 10^{-6}$.

Follow this maths

Follow this maths carefully step by step.

Example 4

Calculate the resistance of **100 m** of **2.5 mm²** aluminium cable, using the resistivity values in Table 2.3.

$$R = \frac{\rho l}{a} \, (\Omega)$$

$$\text{Therefore } R = \frac{28.5 \times 10^{-9} \times 100}{2.5 \times 10^{-6}}$$

$$\therefore R = 1140 \times 10^{-3} \, (\Omega) \text{ or}$$

$$R = 1140 \, (m\Omega)$$

Series connected resistors

When resistors are connected as shown in Fig. 2.4 we say they are connected in series. The same current flows through each resistor and so we say the current is 'common'. When the current flows through R_1 there will be a volt drop across R_1 because of Ohm's law $V = I \times R$. For the same reason a volt drop will occur across R_2 and R_3. The addition of the three volt drops will add up to the total voltage V_T, so:

$$V_T = V_1 + V_2 + V_3 \text{ volts}$$

and from the calculations made in Ohm's law:

$$\text{Total resistance } R_T = R_1 + R_2 + R_3 \text{ ohms}$$

The unit of resistance is the ohm to commemorate the great work done by Dr George Ohm.

Parallel connected resistors

When resistors are connected as shown in Fig. 2.5 we say they are connected in parallel. The same voltage is connected across each resistor and so we say the voltage is common in a parallel circuit. When the current reaches the resistor junction, it will divide, part of it flowing through each resistor. The addition of the three currents will add up to the total current drawn from the battery, so:

$$I_T = I_1 + I_2 + I_3 \text{ amps}$$

and from the calculations made in Ohm's law:

$$\text{Total resistance is found from } \frac{1}{R_T} = \frac{1}{R_1} + \frac{1}{R_2} + \frac{1}{R_3}$$

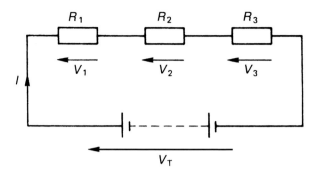

Figure 2.4 A series circuit.

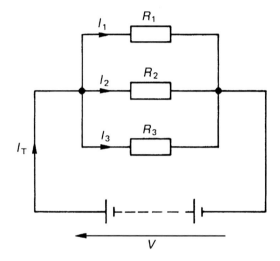

Figure 2.5 A parallel circuit.

Example 5

Three $6\,\Omega$ resistors are connected (a) in series (see Fig. 2.6), and (b) in parallel (see Fig. 2.7), across a 12 V battery. For each method of connection, find the total resistance and the values of all currents and voltages.

For any series connection:

$$R_T = R_1 + R_2 + R_3$$
$$\therefore R_T = 6\,\Omega + 6\,\Omega + 6\,\Omega = 18\,\Omega$$

$$\text{Total current } I_T = \frac{V_T}{R_T}$$

$$\therefore I_T = \frac{12\,V}{18\,\Omega} = 0.67\,A$$

The voltage drop across R_1 is

$$V_1 = I_T \times R_1$$
$$\therefore V_1 = 0.67\,\text{A} \times 6\,\Omega = 4\,\text{V}$$

The voltage drop across R_2 is

$$V_2 = I_T \times R_2$$
$$\therefore V_2 = 0.67\,\text{A} \times 6\,\Omega = 4\,\text{V}$$

The voltage drop across R_3 is

$$V_3 = I_T \times R_3$$
$$\therefore V_3 = 0.67\,\text{A} \times 6\,\Omega = 4\,\text{V}$$

For any parallel connection:

$$\frac{1}{R_T} = \frac{1}{R_1} + \frac{1}{R_2} + \frac{1}{R_3}$$
$$\therefore \frac{1}{R_T} = \frac{1}{6\,\Omega} + \frac{1}{6\,\Omega} + \frac{1}{6\,\Omega}$$
$$\frac{1}{R_T} = \frac{1+1+1}{6\,\Omega} = \frac{3}{6\,\Omega}$$
$$R_T = \frac{6\,\Omega}{3} = 2\,\Omega$$

$$\text{Total current } I_T = \frac{V_T}{R_T}$$
$$\therefore I_T = \frac{12\,\text{V}}{2\,\Omega} = 6\,\text{A}$$

The current flowing through R_1 is

$$I_1 = \frac{V_T}{R_1}$$
$$\therefore I_1 = \frac{12\,\text{V}}{6\,\Omega} = 2\,\text{A}$$

The current flowing through R_2 is

$$I_2 = \frac{V_T}{R_2}$$
$$\therefore I_2 = \frac{12\,\text{V}}{6\,\Omega} = 2\,\text{A}$$

The current flowing through R_3 is

$$I_3 = \frac{V_T}{R_3}$$

$$\therefore I_3 = \frac{12\,V}{6\,\Omega} = 2\,A$$

Try this

Some people find the 'water theory' helpful in understanding series and parallel circuits. So, imagine the resistors to be radiators connected by pipes and the water within the pipe (the current) to be driven by a pump (the voltage).

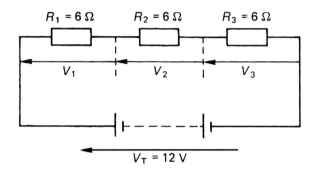

Figure 2.6 Resistors in series.

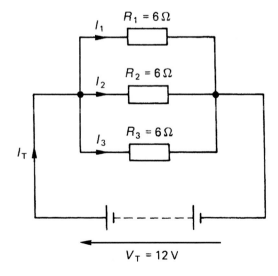

Figure 2.7 Resistors in parallel.

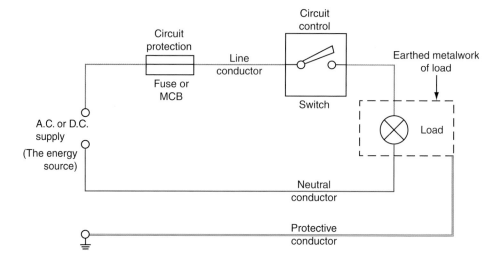

Figure 2.8 Component parts of an electric circuit.

Component parts of an electrical circuit

These series and parallel resistors are connected together to form an electrical circuit. So, what is an electrical circuit?

An electrical circuit has the following **five** components:

- a source of electrical energy. This might be a battery giving a D.C. (direct current) supply or the mains supply which is A.C. (alternating current)
- a source of circuit protection. This might be a fuse or circuit breaker which will protect the circuit from 'overcurrent'
- the circuit conductors or cables. These carry voltage and current to power the load
- a means to control the circuit. This might be a simple on/off switch but it might also be a dimmer or a thermostat
- and a load. This is something which needs electricity to make it work. It might be an electric lamp, an electrical appliance, an electric motor or an iPod.

Connecting voltmeters and ammeters

From the work discussed before we now know that current flows **through** a conductor and voltage appears **across** a resistor, a lamp or any other load. And so, this gives us a good indication of how to connect a voltmeter or ammeter.

An ammeter must have the current flowing through it and so is connected in series with the load. A voltmeter must be connected across the load and so is connected in parallel with the load.

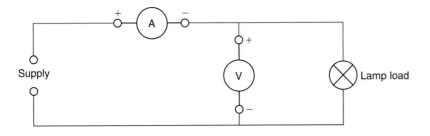

Figure 2.9 Connecting voltmeters and ammeters.

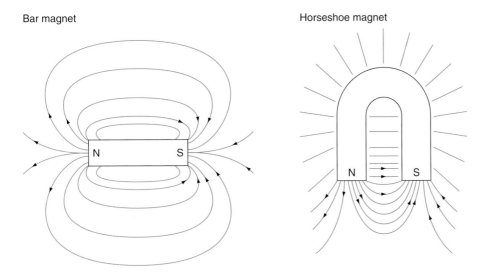

Figure 2.10 Magnetic field around a permanent magnet.

Figure 2.9 shows a voltmeter and ammeter connected to measure the current and voltage in a lamp load.

Magnetic fields and flux patterns

Lines of magnetic flux have no physical existence but were introduced by Michael Faraday as a way of explaining the magnetic energy existing in space or in a material. The magnetic fields around a permanent magnet, a current carrying conductor and a solenoid are shown in Figs 2.10, 2.11 and 2.12.

It is magnetic energy which is used to make commercial electricity today. This was Michael Faraday's great discovery in 1831. Magnetic energy turns electric motors and drives the wheels of industry and is, therefore, important in electrotechnology.

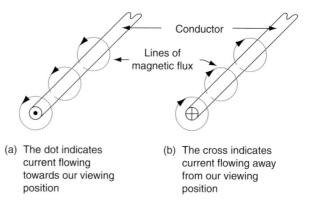

(a) The dot indicates current flowing towards our viewing position

(b) The cross indicates current flowing away from our viewing position

Figure 2.11 Magnetic field around a current carrying conductor.

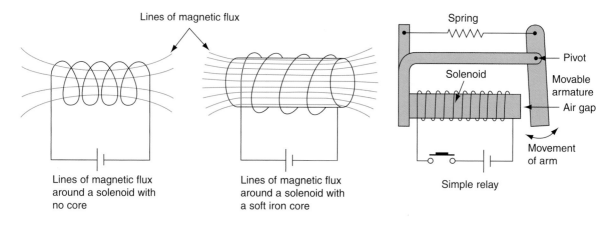

Lines of magnetic flux around a solenoid with no core

Lines of magnetic flux around a solenoid with a soft iron core

Simple relay

Figure 2.12 The solenoid and one practical application, the relay.

Basic mechanics and machines

Mechanics is the scientific study of machines, where a machine may be defined as any device which transmits motion from one place to another. So a lever, a wheel and axle and a pulley are all basic machines. A modern car engine is an energy transforming machine converting fuel energy into motion.

The City & Guilds syllabus asks us to consider weight, mass, force and work done by a force, so let us define some of these scientific terms:

Mass This is a measure of the amount of material in a substance such as wood or metal.

Weight This is a measure of the force which the mass exerts. It exerts this force because it is being attracted towards the earth by gravity.

Force	The presence of a force can only be detected by its effect on an object. A force may cause a stationary object to move or a moving object to stop.
Gravity	The force of gravity acts towards the centre of the earth and causes objects to fall to the ground at a rate of 9.81 m/s.
Work done	The work done by a force is a measure of the force exerted times the distance moved in the direction of the force.

Suppose a broken-down motor car was to be pushed along a road; work would be done on the car by applying the force necessary to move it along the road. Heavy breathing and perspiration would be evidence of the work done:

Work done = Force × Distance moved in the direction of the force (J)

The SI unit of work done is the newton metre or joule (symbol J). The joule is the preferred unit and it commemorates an English physicist, James Prescot Joule (1818–89).

Example 6

A building hoist lifts ten 50 kg bags of cement through a vertical distance of 30 m to the top of a high rise building. Calculate the work done by the hoist, assuming the acceleration due to gravity to be 9.81 m/s^2.

$$\text{Work done} = \text{Force} \times \text{Distance moved (J)}$$
$$\text{but} \quad \text{Force} = \text{Mass} \times \text{Acceleration (N)}$$
$$\therefore \text{Work done} = \text{Mass} \times \text{Acceleration}$$
$$\times \text{Distance moved (J)}$$
$$\text{Work done} = 10 \times 50\,\text{kg} \times 9.81\,\text{m/s}^2 \times 30\,\text{m}$$
$$\text{Work done} = 147.15\,\text{kJ.}$$

Follow this maths

Follow this maths carefully step by step so that you understand where the numbers come from.

Power

If one motor car can cover the distance between two points more quickly than another car, we say that the faster car is more powerful. It can do a given amount of work more quickly. By definition, power is the rate of doing work.

$$\text{Power} = \frac{\text{Work done}}{\text{Time taken}}\,(\text{W})$$

The SI unit of power, both electrical and mechanical, is the watt (symbol W). This commemorates the name of James Watt (1736–1819), the inventor of the steam engine.

Example 7

A building hoist lifts ten 50 kg bags of cement to the top of a 30 m high building. Calculate the rating (power) of the motor to perform this task in 60 seconds if the acceleration due to gravity is taken as 9.81 m/s^2.

$$\text{Power} = \frac{\text{Work done}}{\text{Time taken}} \text{(W)}$$

but   Work done = Force × Distance moved (J)

and   Force = Mass × Acceleration (N)

By substitution,

$$\text{Power} = \frac{\text{Mass} \times \text{Acceleration} \times \text{Distance moved}}{\text{Time taken}} \text{(W)}$$

$$\text{Power} = \frac{10 \times 50 \text{ kg} \times 9.81 \text{ m/s}^2 \times 30 \text{ m}}{60 \text{ s}}$$

$$\text{Power} = 2452.5 \text{ W}$$

The rating of the building hoist motor will be 2.45 kW.

Example 8

A hydroelectric power station pump motor working continuously during a 7 hour period raises 856 tonnes of water through a vertical distance of 60 m. Determine the rating (power) of the motor, assuming the acceleration due to gravity is 9.81 m/s^2.

From Example 7,

$$\text{Power} = \frac{\text{Mass} \times \text{Acceleration} \times \text{Distance moved}}{\text{Time taken}} \text{(W)}$$

$$\text{Power} = \frac{856 \times 1000 \text{ kg} \times 9.81 \text{ m/s}^2 \times 60 \text{m}}{7 \times 60 \times 60 \text{s}}$$

$$\text{Power} = 20\,000 \text{ W}$$

The rating of the pump motor is 20 kW.

Example 9

An electric hoist motor raises a load of 500 kg at a velocity of 2 m/s. Calculate the rating (power) of the motor if the acceleration due to gravity is 9.81 m/s^2.

$$\text{Power} = \frac{\text{Mass} \times \text{Acceleration} \times \text{Distance moved}}{\text{Time taken}} \text{(W)}$$

$$\text{but} \quad \text{Velocity} = \frac{\text{Distance}}{\text{Time}} \text{(m/s)}$$

$$\therefore \text{Power} = \text{Mass} \times \text{Acceleration} \times \text{Velocity}$$

$$\text{Power} = 500\,\text{kg} \times 9.81\,\text{m/s}^2 \times 2\,\text{m/s}$$

$$\text{Power} = 9810\text{ W.}$$

The rating of the hoist motor is 9.81 kW.

Follow this maths

$$\text{Power} = \frac{\text{Work done}}{\text{Time taken}}$$

But, Work done = Force × Distance moved

And, Force = Mass × Acceleration

Therefore by substitution:

$$\text{Power} = \frac{\text{Mass} \times \text{Acceleration} \times \text{Distance moved}}{\text{Time taken}}$$

Efficiency

In any machine the power available at the output is less than that which is put in because losses occur in the machine. The losses may result from friction in the bearings, wind resistance to moving parts, heat, noise or vibration.

The ratio of the output power to the input power is known as the *efficiency* of the machine. The symbol for efficiency is the Greek letter 'eta' (η). In general:

$$\eta = \frac{\text{Power output}}{\text{Power input}}$$

Since efficiency is usually expressed as a percentage we modify the general formula as follows:

$$\eta = \frac{\text{Power output}}{\text{Power input}} \times 100$$

Example 10

A transformer feeds the 9.81 kW motor driving the mechanical hoist of the previous example. The input power to the transformer was found to be 10.9 kW. Find the efficiency of the transformer.

$$\eta = \frac{\text{Power output}}{\text{Power input}} \times 100$$

$$\eta = \frac{9.81\text{ kW}}{10.9\text{ kW}} \times 100 = 90\%$$

Thus the transformer is 90% efficient. Note that efficiency has no units, but is simply expressed as a percentage.

The simple alternator

If a loop of wire is rotated between the poles of a magnet as shown in Fig. 2.13, the loop of wire will cut the lines of magnetic flux which pass from the north to the south pole. This flux cutting causes a voltage to be induced in the loop of wire (Faraday's Law).

If this induced voltage is collected by carbon brushes at the slip rings and displayed on a meter or cathode ray oscilloscope (CRO), it will be seen to induce first a positive and then a negative voltage. We call this changing voltage an alternating voltage and the shape that it follows is called, in mathematics, sinusoidal.

Electrical transformers

A transformer is an electrical machine without moving parts, which is used to change the value of an alternating voltage.

www

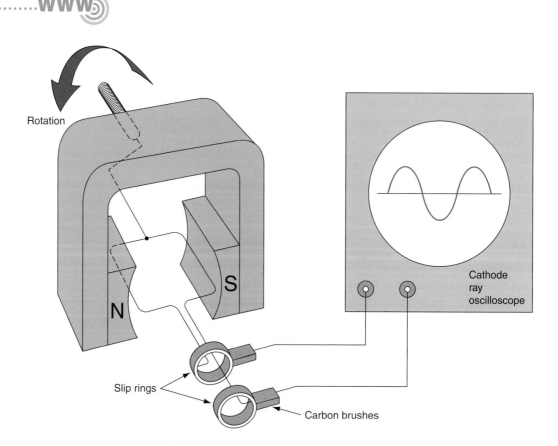

Figure 2.13 Simple A.C. generator or alternator.

A transformer will only work on an alternating supply; it will not normally work from a D.C. supply such as a battery.

- A transformer such as that shown in Fig. 2.14 consists of two coils called the primary and secondary coils or windings, wound on to a common core. The iron core of the transformer is not solid but made up of very thin sheets called laminations, to improve efficiency.
- An alternating voltage applied to the primary winding establishes an alternating magnetic flux in the core.
- The magnetic flux in the core causes a voltage to be induced in the secondary winding of the transformers.
- The voltage in both the primary and secondary windings is proportional to the number of turns.
- This means that if you increase the number of secondary turns you will increase the output voltage. This has an application in power distribution.
- Alternatively, reducing the number of secondary turns will reduce the output voltage. This is useful for low voltage supplies such as domestic bell transformers. Because it has no moving parts, a transformer can have a very high efficiency. Large power transformers, used on electrical distribution systems, can have an efficiency of better than 90%.

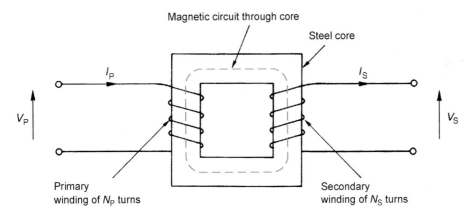

Figure 2.14 A simple transformer.

Large power transformers need cooling to take the heat generated by the losses away from the core. This is often achieved by totally immersing the core and windings in insulating oil. A sketch of an oil immersed transformer can be seen in Fig. 2.15.

Very small transformers are used in electronic applications. Small transformers are used as isolating transformers in shaver sockets and can also be used to supply SELV (separated extra low voltage) sources. Equipment supplied from a SELV source may be installed in a bathroom or shower room, provided that it is suitably enclosed and

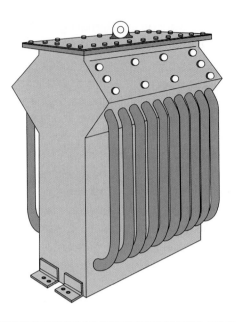

Figure 2.15 Typical oil filled power transformer.

protected from the ingress of moisture. This includes equipment such as water heaters, pumps for showers and whirlpool baths.

Try this

Have you seen any transformers in action?

Were they big or small – what were they being used for?

I take my students to Heysham Power Station – have you been to a power station? Power stations are very large installations.

Have you been close up to a transmission tower, perhaps when you were walking in the countryside?

Be observant, look around you. Those of us in the electrotechnical industry see lots of electrical details that others do not see.

Electrical power on the national grid

Electricity is generated in large modern power stations at 25 kV (25 000 volts). It is then transformed up to 132 kV or 270 kV for transmission to other parts of the country on the national grid network. This is a network of overhead conductors suspended on transmission towers which link together the power stations and the millions of users of electricity.

Raising the voltage to these very high values reduces the losses on the transmission network. 66 kV or 33 kV are used for secondary transmission lines and then these high voltages are reduced to 11 kV

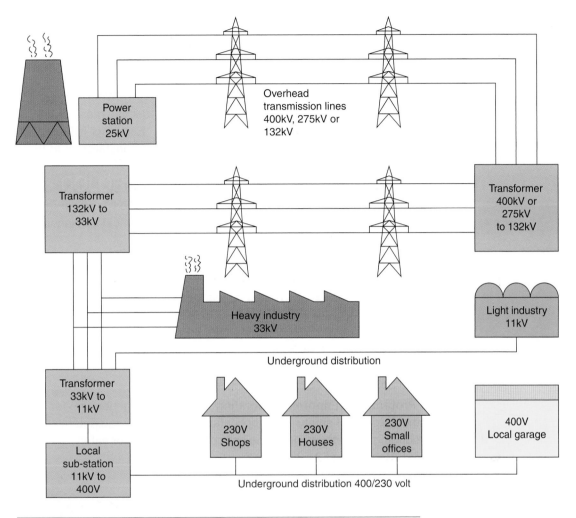

Figure 2.16 Simplified diagram of the distribution of electricity from power station to consumer.

at local sub-stations for distribution to end-users such as factories, shops and houses at 400 V and 230 V.

The ease and efficiency of changing the voltage levels is only possible because we generate an A.C. supply. Transformers are then used to change the voltage levels to those which are appropriate. Very high voltages for transmission, lower voltages for safe end use. This would not be possible if a D.C. supply was generated.

Figure 2.16 shows a simplified diagram of electricity distribution.

Safe electrical systems

Installing electrical systems which will be safe for those who will use them is absolutely fundamental to the safe use of electricity. Electrical

Key fact

You should remember all of the definitions of words in this section. Perhaps you could make a list of the words and write down a short definition for each one. You will probably need to read this section more than once before you understand it, but it is important.

systems installed in accordance with the IEE Regulations (BS 7671) will be safe for those who will use them.

Chapter 13 of the IEE Regulations tells us that where electrical equipment may become charged with electricity so as to cause a danger, any metalwork must be connected to earth. When we say connected to '**earth**' we mean the general conductive mass of the planet Earth, whose potential is taken as zero. '**Earthing**' is the act of connecting the '**exposed conductive parts**' of an installation to the main earthing terminal of the installation.

'**Exposed conductive parts**' are the metal parts of the installation which are **not normally live** but which may become live under fault conditions. For example, the metalwork of an electrical appliance or the trunking and conduits of the installation.

All other metalwork within a building is called '**extraneous conductive parts**' and this includes structural steelwork and other service pipes such as gas, water, radiators and sinks. The extraneous conductive parts are prevented from becoming live by '**bonding**' them together and connecting them to the main earthing terminal of the installation. The bonding process maintains an '**equipotential**' (of zero volts) between all exposed and extraneous conductive parts.

'**Protective equipotential bonding**' is equipotential bonding for the purpose of safety. The application of protective equipotential bonding to earth is one of the important principles for safety. (IEE Regulation 131.2.2).

Principles of electric shock protection

An electric shock occurs when a person becomes a part of the electrical circuit. We looked at electric shock in Chapter 1 at Fig. 1.8. The intensity of the electric shock will depend upon many factors such as age, fitness and the circumstances in which the shock is received. In general terms, a shock current of more than 50 mA can be fatal.

Electric shock may occur in two ways, through direct contact or indirect contact with live parts. Protection against actually touching live parts is provided by:

- insulating live parts
- placing barriers or enclosures around live parts
- placing obstacles in front of live parts
- placing live parts out of reach.

Each of these methods keep people away from live electrical equipment and is called **Basic Protection** (IEE Regulation 131.2.1).

Indirect contact means touching exposed conductive parts, such as the metalwork of an appliance, which has become live as a result of a fault. The potential voltage on this metalwork rises above earth potential and an electric shock may occur when someone touches the metalwork.

Protection against touching something made live as a result of a fault, is called **Fault Protection** and is achieved by protective equipotential bonding and automatic disconnection of the supply in the event of a fault occurring (IEE Regulation 131.2.2).

Protective equipotential bonding coupled with automatic disconnection of the supply

In the UK the most universally used method of fault protection is protective equipotential bonding coupled with automatic disconnection of the electrical supply. Protective equipotential bonding was discussed at the beginning of this section and is the process of connecting all exposed conductive parts and extraneous conductive parts to the main earthing terminal of the electrical installation. Automatic disconnection of the supply is achieved by fuses, MCBs and RCDs.

If the circuit shown earlier in Figure 2.8 was operating normally, current would flow from the supply to the load along the line conductor, through the load and back along the neutral conductor. The protective device would be chosen to carry this current. However, if a fault occurs, for example a short circuit to earth between the line conductor and the earthed metalwork of the load, current will flow from the supply to the load and then through the low resistance earthing and bonding of the installation back to the supply. This will cause a large current to flow and, in a healthy circuit, the protective device will operate very quickly to remove the danger.

Fuses, MCBs and RCDs provide earth fault protection, overload protection and short circuit protection where:

- a short circuit is a fault of negligible impedance (call it resistance for now) between live and neutral conductors
- an overload is a current which exceeds the rated value in an otherwise healthy circuit.

In all cases the basic requirement for protection is that the fault current should be removed quickly and the circuit isolated. The IEE Regulations state that the protective device must operate very quickly to remove the danger.

IEE Regulation 411.3.2 tells us that for final circuits not exceeding 32 A, the maximum disconnection time shall not exceed 0.4 seconds.

> **Key fact**
>
> Protective equipotential bonding
>
> - bonding clamps must be of an approved type
> - must be fitted to cleaned pipework
> - must be tight and secure
> - must have a visible label
> - IEE Regulation 514.13.1.

Electrical tools and equipment

Good quality, sharp tools are important to any craftsman; they enable learned skills to be used to the best advantage. The basic tools required by anyone in the electrotechnical industry are those used for stripping and connecting conductors. These are pliers, side cutters, a knife and an assortment of screwdrivers with flat bladed, Phillips, crosshead, Pozidriv, Torx or Hexidriv bits. Figure 2.17 shows the basic hand tools required for making electrical connections.

The additional tools required by an electrical craftsman will depend upon the type of electrotechnical work being undertaken. When wiring new houses or re-wiring old ones, the additional tools are those more associated with a bricklayer or carpenter and some examples are shown in Fig. 2.18.

When working on industrial installations, installing conduit, trunking and tray, the additional tools required by an electrician would more normally be those associated with a fitter or sheet metal fabricator and some examples are shown in Fig. 2.19.

The special tools required for cable tray bending, steel conduit bending and screw threading stocks and dies, plus M.I. Cable crimping tools are shown in Fig. 2.20.

Electrical power tools reduce much of the hard work for any craftsman, allowing an increase in productivity. Battery powered tools

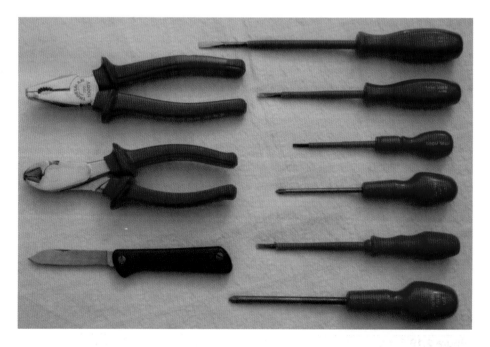

Figure 2.17 The tools used for making electrical connections.

are very popular because they are very safe to use on site and are often now supplied with two battery packs so that while one is being used the other is on charge. Only 110 V power tools with leads are allowed on most construction sites these days. Figure 2.21 shows a selection of electrical power tools.

Figure 2.18 Some additional tools required by an electrician engaged in house wiring.

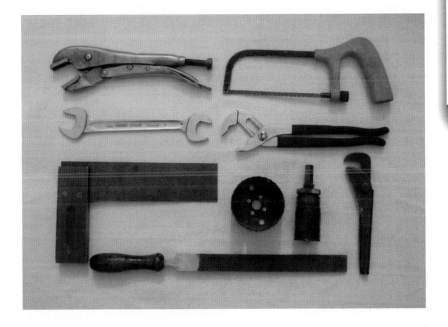

Figure 2.19 Some additional tools required by an electrician engaged in industrial installations.

Key facts

Safety rules for hand tools

- always use the correct tool for the job in hand and use it properly and sensibly
- always keep tools clean and sharp
- always keep tools in a toolbox and secure

Safety rules for power tools

- always check that the casing is not damaged
- always check that the cable is not damaged
- always check that the plug top is not damaged
- always check that no coloured conductors are showing anywhere on the flexible cord
- always check that mains power tools have been properly tested (PAT tested) and carry a label

Finally, and most importantly

- If it is broken or damaged in any way DO NOT USE IT
- Ask a 'competent person' (probably your supervisor) to check it out
- Do not let anyone else use it

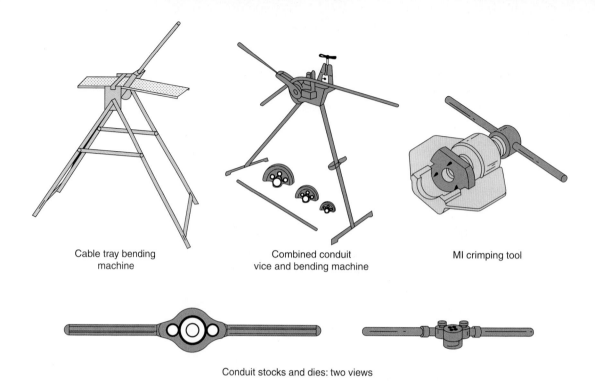

Cable tray bending
machine

Combined conduit
vice and bending machine

MI crimping tool

Conduit stocks and dies: two views

Figure 2.20 Some special tools required by an electrician engaged in industrial installations.

Figure 2.21 Electrical power tools.

Before using any power tools, the craftsman should inspect the tool and any associated flexible cords for damage. If the power tool carries a PAT (portable appliance testing) label, a check should be made to ensure that the test date has not expired.

All tools are expensive and, therefore, attractive to a thief so, when not in use, all tools must be stored safely and securely.

Safe working practice

Every year thousands of people have accidents at their place of work despite the legal requirements laid down by the Health & Safety Executive. Many people recover quickly but an accident at work can result in permanent harm or even death.

At the very least, injuries hurt individuals. They may prevent you from doing the things you enjoy in your spare time and the result could be loss of earnings to you and loss of production, and possibly damage to equipment, for your employer. Your place of work may look harmless but it can be dangerous.

You have a responsibility under the Health & Safety at Work Act to:

- learn how to work safely and to follow company procedures of work
- obey all safety rules, notices and signs
- not interfere with or misuse anything provided for safety
- report anything that seems damaged, faulty or dangerous
- behave sensibly, not play practical jokes and not distract other people at work
- walk sensibly and not run around the workplace
- use the prescribed walkways
- drive only those vehicles for which you have been properly trained and passed the necessary test
- not wear jewellery which could become caught in moving parts if you are using machinery at work
- always wear appropriate clothing and PPE if necessary.

The principles laid down in the many health & safety at work regulations control our working environment and make our workplaces safer but despite all the legislation, workers continue to be injured and killed at work.

The Health & Safety Executive (HSE) statistics show that more than 200 people die each year as a result of a work related injury. In addition, about 28 000 people have serious injuries each year and about 130 000 people each year receive minor work related injuries which result in an absence from work for more than three days.

Figure 2.22 Safe manual handling.

The most common causes of accidents at work are:

- slips, trips and falls from above ground. Safe working above ground is discussed at the beginning of Chapter 3
- manual handling, that is moving objects by hand which may result in strains, sprains and trap injury pains. Always use a mechanical aid to move heavy objects. Safe manual handling is discussed at the beginning of Chapter 3
- using equipment, machines and tools. Make sure your tools and equipment meet the safety rules described in the last section
- storing equipment badly which then becomes unstable and falls on someone
- fire – we discussed fire safety in Chapter 1
- electricity – the safe use of electricity is what this book and our industry is about. Always use the 'safe isolation procedure' before work begins as described in Chapter 1.

To help prevent accidents at work:

- always behave sensibly and responsibly
- keep your work area clean and tidy
- keep walkways clear
- clean up spills or wet patches on the floor
- screen off your work areas from the general public and other trades
- put tools and equipment away when not in use. Do not leave things lying around for others to fall over
- when working above ground level, use the good practice described in the next chapter
- when moving objects by hand, use the good practice described in the next chapter under the heading 'Safe manual handling'.

Assessment questions

Identify the statements as true or false. If only part of the statement is false, tick false.

1   The electrical properties of any material basically depend upon how tightly the electrons are attached to the nucleus of the atom. A strong bond, and the material will be an insulator, a weak bond and the material will be a good conductor of electricity.

   True ☐   False ☐

2   An electrical current flowing in a conductor will have a heating, magnetic or chemical effect upon the circuit. Increasing the current flow will always **reduce** one of the three effects.

   True ☐   False ☐

3   The pioneering work carried out by Dr George Ohm in 1826 allows us today to calculate the relationship between current, voltage and resistance in an electric circuit.

   True ☐   False ☐

4   Measuring the current and voltage in an electric circuit is an important practical skill for anyone in the electrotechnical industry. When measuring current, the ammeter is always connected across the load. When measuring voltage, the voltmeter is always connected in series with the load.

   True ☐   False ☐

5   Electrical cables are used to carry electrical currents. Most cables are constructed in three parts:
   - the conductor, which prevents human beings and livestock from an electric shock
   - the insulation, which carries the electric current, thicker cables carry more current
   - the outer sheath, which may incorporate a means of protection from mechanical damage

   True ☐   False ☐

6 Good quality sharp tools are important to a craftsman in the electrotechnical industry. Always apply the following safety rules to both hand and power tools:
   - keep tools sharp, clean and in a toolbox when not in use
   - always use the correct tools and leads
   - check for damage to power tools and leads
   - mains power tools should be PAT tested and carry a label which is 'in date'
   - if it's broken or damaged, don't use it, and don't let anyone else use it

   True   False ☐

7 Slips, trips and falls are the most common cause of accidents in the workplace. To prevent accidents at work always:
   - behave sensibly and don't fool about
   - keep your work area clean and avoid tripping hazards
   - when working above ground, work from a suitable platform
   - when lifting objects by hand, use the 'safe manual handling technique'

   True ☐   False ☐

8 When those of us who work in the electrotechnical industry use the phrase 'to connect to earth', we mean connect to the main earthing terminal of the electrical installation so as to make it safe

   True ☐   False ☐

9 The metal trunking and conduits of an electrical installation are called the **extraneous conductive parts**

   True ☐   False ☐

10 Structural steelwork, metal service pipes and heating radiators are called **exposed conductive parts**

   True ☐   False ☐

Multiple choice assessment questions

Tick the correct answer. Note that more than ONE answer may be correct.

11 In the SI system of units, the units of voltage, current and resistance are:

a. volts, watts and newtons ☐

b. metre, kilogram and second ☐

c. volts, amps and ohms ☐

d. newton, joule and watt ☐

12 In the SI system of units, the units of length, mass and time are:

a. volts, watts and newtons ☐

b. metre, kilogram and second ☐

c. volts, amps and ohms ☐

d. newton, joule and watt ☐

13 In the SI system of units, the units of force, energy and power are:

a. volts, watts and newtons ☐

b. metre, kilogram and second ☐

c. volts, amps and ohms ☐

d. newton, joule and watt ☐

14 Electricity is generated in power stations at 25 kV. In the SI system of units 25 kV may be written as:

a. 25 volts or 25 thousand volts ☐

b. 25×10^{-3} amps or $25 \div 1000$ amps ☐

c. 25×10^{3} volts or 25 000 volts ☐

d. 25 amps or 25 thousand amps ☐

15 Electronic equipment uses very small amounts of current. In the SI system of units 25 milliamperes may be written as:

a. 25 volts or 25 thousand volts ☐

b. 25×10^{-3} amps or $25 \div 1000$ amps ☐

c. 25×10^{3} volts or 25 000 volts ☐

d. 25 amps or 25 thousand amps ☐

16 An insulator is a material in which the electrons are:

a. very large compared with the nucleus ☐

b. positively charged to the nucleus ☐

c. tightly bound to the nucleus ☐

d. loosely bound to the nucleus ☐

17 A conductor is a material in which the electrons are:

a. very large compared with the nucleus ☐

b. positively charged to the nucleus ☐

c. tightly bound to the nucleus ☐

d. loosely bound to the nucleus ☐

18 A 'good conductor' material has:

a. a negative nucleus in the atoms of the material ☐

b. positive electrons available for current flow ☐

c. free electrons available for current flow ☐

d. no free electrons ☐

19 A 'good insulator' material has:

a. a negative nucleus in the atoms of the material ☐

b. positive electrons available for current flow ☐

c. free electrons available for current flow ☐

d. no free electrons ☐

20 The following materials are good conductors:

a. copper, perspex and glass ☐

b. copper, brass and wood ☐

c. copper, brass and aluminium ☐

d. PVC, rubber and porcelain ☐

21 The following materials are good insulators:

a. PVC, copper and aluminium ☐

b. PVC, rubber and brass ☐

c. PVC, rubber and porcelain ☐

d. copper, gold and silver ☐

22 An electric current in a circuit may also be described as a:

a. flow of atoms ☐

b. difference of potential ☐

c. resistance in the circuit ☐

d. flow of free electrons ☐

23 Most electrical cables are constructed in three parts, the:

a. conductor, copper and aluminium ☐

b. conductor, insulation and flexible cord ☐

c. conductor, insulation and sheath ☐

d. conductor, outer sheath and protection ☐

24 PVC insulated and sheathed cables and cords would be suitable for the following situations:

a. the fixed wiring in domestic installations ☐

b. the fixed wiring in industrial installations ☐

c. flexible cords connecting domestic appliances to a 13A socket outlet ☐

d. an underground cable to a remote building such as a domestic garage ☐

25 PVC/SWA cables would be suitable for the following situations:

a. the fixed wiring in domestic installations ☐

b. the fixed wiring in industrial installations ☐

c. flexible cords connecting domestic appliances to a 13A socket outlet ☐

d. an underground cable to a remote building such as a domestic garage ☐

26 When an electric current flows in an electric circuit it can have one or more of the following three effects:

a. voltage, resistance and current ☐

b. steaming, smoking and getting hot ☐

c. heating, magnetic and chemical ☐

d. conduction, convection and radiation ☐

27 Using Ohm's Law, calculate the resistance of a circuit in which the voltage was 230V and the current 5A:

a. 21.7 ohm ☐

b. 46.0 ohm ☐

c. 460 ohm ☐

d. 1150 ohm ☐

28 Using Ohm's Law, calculate the current flowing in a 230V kettle element of resistance 19.166 ohm:

a. 8.33A ☐

b. 12.00A ☐

    c.   16.66 A   ☐

    d.   4408 A   ☐

29 **Using Ohm's Law, calculate the voltage connected across a resistor of 1000 ohm when a current of 3 milliamperes flows:**

    a.   3 mV   ☐

    b.   3 V   ☐

    c.   30 V   ☐

    d.   300 V   ☐

30 **Calculate the resistance of a one metre bar of silver, 1.5 mm² in cross-sectional area if the resistivity of silver is $16.4 \times 10^{-9}\,(\Omega\text{m})$:**

    a. 10.93×10^{-3} ohm   ☐

    b.   10.93 milli-ohm   ☐

    c. 91.46×10^{-3} ohm   ☐

    d.   91.46 milli-ohm   ☐

31 **Calculate the resistance of a one metre bar of iron, 1.5 mm² in cross-sectional area, if the resistivity of iron is $100 \times 10^{-9}\,(\Omega\text{m})$:**

    a. 15.00×10^{-3} ohm   ☐

    b. 66.66×10^{-3} ohm   ☐

    c.   66.66 milli-ohm   ☐

    d.   15.00 milli-ohm   ☐

32 **The resistance of the iron bar in Question 31 above, compared with the resistance of the silver bar in Question 30 is:**

    a.   the iron bar has about 6 times less resistance   ☐

    b.   the iron bar has about 6 times more resistance   ☐

    c.   the iron bar has about 15 times less resistance   ☐

    d.   the iron bar has about 15 times more resistance   ☐

33 **Two 6 ohm resistors are connected first in series and then in parallel. For each connection calculate the total resistance:**

    a.   series 2 ohm parallel 3 ohm   ☐

    b.   series 3 ohm parallel 2 ohm   ☐

    c.   series 3 ohm parallel 12 ohm   ☐

    d.   series 12 ohm parallel 3 ohm   ☐

34 **Three resistors of 24, 40 and 60 ohms are connected first in series and then in parallel. For each connection calculate the total resistance:**

    a.   series 124 ohm parallel 2.4 ohm   ☐

    b.   series 124 ohm parallel 12 ohm   ☐

    c.   series 124 ohm parallel 15 ohm   ☐

    d.   series 124 ohm parallel 124 ohm   ☐

35 Three 2 ohm resistors are connected first in series and then in parallel across a 12 volt battery supply. Calculate the current flowing for each connection:

   a. series 2 A parallel 18.18 A ☐

   b. series 8 A parallel 12.5 A ☐

   c. series 12.5 A parallel 8.0 A ☐

   d. series 18.18 A parallel 2.0 A ☐

36 To correctly measure the current and voltage in a circuit, the meters must be connected to the load in the following way:

   a. ammeter in series, voltmeter across the load ☐

   b. ammeter across the load, voltmeter in series ☐

   c. ammeter in series, voltmeter in parallel ☐

   d. ammeter in parallel, voltmeter in series ☐

37 Magnetic energy causes:

   a. like poles to attract ☐

   b. unlike poles to repel ☐

   c. like poles to repel ☐

   d. unlike poles to attract ☐

38 A measure of the amount of material in a substance is called its:

   a. force ☐

   b. gravity ☐

   c. mass ☐

   d. weight ☐

39 The force which acts toward the centre of the earth is called:

   a. force ☐

   b. gravity ☐

   c. mass ☐

   d. weight ☐

40 Calculate the work done (WD) by a 50 kg bag of cement when it falls 10 metres from a scaffold to the ground. What do you think might be the consequences of this action? What might be the consequences if the bag fell on to a worker below? Assume the acceleration due to gravity to be 9.81 m/s^2:

   a. WD 3.996 kJ – bag remains intact ☐

   b. WD 4.90 kJ – bag bursts open ☐

   c. WD 4.90 kJ – bag remains intact ☐

   d. WD 50.96 kJ – bag bursts open ☐

The worker below would certainly be injured, possibly seriously.

41 Calculate the efficiency of a 1 kW electric motor which takes 1200 W from the source of supply:

   a.   10.9%   ☐
   b.   12.43%   ☐
   c.   83.33%   ☐
   d.   120%   ☐

42 Increasing the number of secondary turns on a transformer connected to an A.C. supply will:

   a.   decrease the input voltage   ☐
   b.   decrease the output voltage   ☐
   c.   increase the input voltage   ☐
   d.   increase the output voltage   ☐

43 The iron core of a transformer is:

   a.   solid so as to increase the core magnetic flux   ☐
   b.   laminated so as to increase the core magnetic flux   ☐
   c.   solid in order to reduce the losses   ☐
   d.   laminated in order to reduce the losses   ☐

44 The metal parts of a building structure are called:

   a.   earthing   ☐
   b.   equipotential bonding   ☐
   c.   exposed conductive parts   ☐
   d.   extraneous conductive parts   ☐

45 The metal parts of an electrical installation not normally live are called:

   a.   earthing   ☐
   b.   equipotential bonding   ☐
   c.   exposed conductive parts   ☐
   d.   extraneous conductive parts   ☐

46 The act of connecting exposed conductive parts to the earthing terminal of an installation is called:

   a.   earthing   ☐
   b.   equipotential bonding   ☐
   c.   exposed conductive parts   ☐
   d.   extraneous conductive parts   ☐

47 The process which maintains a potential of zero volts between all exposed and extraneous parts is called:

   a.   earthing   ☐
   b.   equipotential bonding   ☐
   c.   exposed conductive parts   ☐
   d.   extraneous conductive parts   ☐

48 **Pliers, cutters, a knife and a range of screwdrivers are the tools required in the electrotechnical industry for:**

   a.   erecting conduit ☐

   b.   assembling tray ☐

   c.   stripping and connecting conductors ☐

   d.   terminating an MI cable ☐

49 **When visually inspecting an electrical power tool before using it, you notice minor damage to the case and the coloured conductors showing at the junction with the plug top. This power tool, in this condition should:**

   a.   not be used by the company trainee ☐

   b.   only be used if the PAT test label is 'in date' ☐

   c.   only be used by a 'competent person' ☐

   d.   not be used until inspected and tested by a 'competent person' ☐

50 **The most common cause of accidents at work is:**

   a.   gloves, boots and hard hats ☐

   b.   sprains, strains and trap pains ☐

   c.   slips, trips and falls ☐

   d.   hook, line and sinker. ☐

CHAPTER 3

Health and safety application and electrical principles

Online Material WWW

To support your learning, animated versions of some of the figures from this chapter are viewable for free from the book's companion site. **When you see the logo, visit the website below to access them:**
www.elsevierdirect.com/
companions/9780080969404

This chapter describes safe systems of working and the principle of operation of some electrical machines, equipment and systems.

Health and safety applications

Avoiding accidents in the workplace

The Health & Safety at Work Act 1974 places a statutory and common law obligation on employers to take reasonable care of the health and safety of their workers. The Management of Health & Safety at Work Regulations 1999 place an obligation on employers to carry out 'risk assessments' and, where necessary, to take action to eliminate or control risks. The Workplace (Health, Safety and Welfare) Regulations 1992 and the Construction (Health, Safety and Welfare) Regulations 1996 cover all aspects of the workplace and construction sites respectively. They include the requirement that all areas where people could fall from a height of two metres or above, are properly guarded. The Work at Height Regulations were introduced in April 2005. The aim of these regulations is to avoid working at height, if possible, but where this cannot be avoided, to use the best practicable means of ensuring the safety of those working at height. However, despite all the legislation, we know from the HSE statistics that accidents still occur in the workplace.

3

Figure 3.1 Slips, trips and falls are the most common causes of accidents in the workplace.

The most common causes of accidents in the workplace are:

- slips, trips and falls
- manual handling, that is moving objects by hand
- using equipment, machinery or tools
- storage of goods and materials which then become unstable and fall on someone
- fire
- electricity
- mechanical handling

To control the risk of an accident we usually:

- eliminate the cause; that means, do not do the job or procedure in an unsafe way
- substitute a procedure or product with less risk; that means finding a safer way to complete the job or procedure
- enclose the dangerous situation; that means fitting guards or screening off an area and only allowing trained and competent people into a potentially dangerous area
- put guards around a hazard; for example, placing guards in front of cutting and grinding wheels
- use safe systems of work; that means establishing written procedures for work that is potentially dangerous. These written procedures are sometimes called 'permits to work' or 'method statements'
- supervise, train and give information to staff which leads to a 'competent' workforce
- if a hazard cannot be removed or minimized, then the employer must provide PPE. However, providing personal protective equipment to staff must be a last resort when the hazard cannot be removed in any other way. The PPE must be provided at the employer's expense

A **hazard** is something with the **potential to cause harm**; for example, electric tools, working above ground level, wet or uneven floors, rotating parts.

A **risk** is the **possibility of harm actually being done**. Is it a high, low or medium risk? Who is at risk—the office staff, electricians, the public? Is the risk adequately controlled?

A **positive, personal attitude to safety** reduces accidents at work. Always work and act responsibly and safely to protect yourself and others. Be aware of the hazards around you, the protection available to you and the means of preventing accidents.

Risk assessment, the process

We have already said that an employer must carry out risk assessments as a part of a robust health and safety policy. The HSE recommends five steps to any risk assessment.

Step 1

Look at what might reasonably be expected to cause harm. Ignore the trivial and concentrate only on significant hazards that could result in serious harm or injury. For example:

- Slipping, tripping or falling hazards, e.g. from poorly maintained or partly installed floors and stairs
- Fire, e.g. from flammable materials you might be using such as solvents
- Rotating parts of hand tools, e.g. drills
- Accidental discharge of cartridge operated tools
- Manual handling, e.g. lifting, moving or supporting loads

Step 2

Decide who might be harmed. Do not list individuals by name. Just think about groups of people doing similar work or who might be affected by your work:

- Office staff
- Electricians
- Maintenance personnel
- Other contractors on site

Step 3

Evaluate what is the risk arising from an identified hazard. Is it adequately controlled or should more be done? Is the risk low, medium or high? Only low risk will be acceptable when the HSE Inspector comes to inspect your company records. Do the precautions already taken:

- meet the legal standards required
- comply with recognized industrial practice
- represent good practice
- reduce the risk as far as is reasonably practicable

If you can answer 'yes' to the above points then the risks are adequately controlled, but you need to state the precautions that have been put in place, e.g. electric shock hazard from using portable equipment is reduced to low by PAT testing all equipment every 6 months.

HAZARD RISK ASSESSMENT	FLASH-BANG ELECTRICAL CO.
For Company name or site: _____ Address: _____ _____	Assessment undertaken by: _____ Signed: _____ Date: _____

STEP 5   Assessment review date: _____

STEP 1 List the hazards here	STEP 2 Decide who might be harmed
------------------------------------ ------------------------------------ ------------------------------------ ------------------------------------ ------------------------------------ ------------------------------------ ------------------------------------ ------------------------------------ ------------------------------------ ------------------------------------ ------------------------------------	------------------------------------ ------------------------------------ ------------------------------------ ------------------------------------ ------------------------------------ ------------------------------------ ------------------------------------ ------------------------------------ ------------------------------------ ------------------------------------
STEP 3 Evaluate (what is) the risk – is it adequately controlled? State risk level as low, medium or high	STEP 4 Further action – what else is required to control any risk identified as medium or high?
------------------------------------ ------------------------------------ ------------------------------------ ------------------------------------ ------------------------------------ ------------------------------------ ------------------------------------ ------------------------------------	------------------------------------ ------------------------------------ ------------------------------------ ------------------------------------ ------------------------------------ ------------------------------------ ------------------------------------ ------------------------------------

Figure 3.2 Hazard risk assessment – standard form.

Step 4

Further action – what more could be done to reduce those risks which were found to be inadequately controlled?

Any hazard identified by a risk assessment as high risk must be brought to the attention of the person responsible for health and safety within the company.

Step 5

The assessment must be reviewed from time to time by the person responsible for health and safety.

Try this

To help you to be more aware of the hazards around you at work, you might like to carry out a risk assessment on a situation you are familiar with at work, using the standard form of Fig. 3.2 or your employer's standard form. Make a few photocopies and ask your supervisor to help you, perhaps one lunch time.

Safe manual handling

There have been so many injuries over the years as a result of lifting, transporting or supporting loads by hand or bodily force that the Health & Safety Executive has introduced legislation, the Manual Handling Operations Regulations 1992. These state that:

- if a job involves considerable manual handling, workers must be trained in the correct lifting procedure
- loads must not be lifted manually if it is more appropriate to use a mechanical aid
- always use a trolley, sack truck or wheelbarrow when these are available
- use good manual lifting techniques if the load must be lifted manually and avoid jerky movements
- only lift and carry what you can manage easily
- wear gloves to avoid rough or sharp edges.

Good manual lifting techniques

When manually lifting objects from the floor:

- bend at the hips and knees to get down to the object
- grasp the object firmly
- take account of its centre of gravity

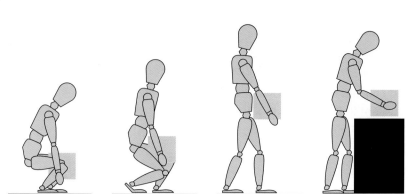

Figure 3.3 Correct manual lifting and carrying procedure.

- keep your back straight and head erect, use the powerful leg muscles to raise the object
- carry the load close to the body.

Safe working above ground level

Working above ground level is hazardous because there is a risk of falling. If the working platform is appropriate for the purpose, properly erected and in good condition, then the risk is low.

However, in 2004 the HSE statistics show that there were 67 fatal falls and almost 4000 major injuries resulting from falls. They are the biggest single cause of workplace deaths and one of the main causes of major injury. If you fall from a height above two metres, statistically you are 'very likely' to sustain a serious injury. To reduce accidents as a result of falls from height the HSE introduced The Work at Height Regulations 2005. They became law in April 2005.

The main hazards associated with working at height are people falling and objects falling on to people. The main aim of the regulations is to:

- avoid working at height if possible
- no work must be done at height if it is safe and reasonably practicable to do it other than at height
- use work equipment to prevent falls where it is impossible to avoid working at height. That is guard rails and toe boards on scaffold platforms
- where the risk of a fall cannot be eliminated, robust platforms must be built so as to 'minimize' the distance and consequences should a fall occur. This may mean building a stepped pyramid type of platform at the work site when the risk is high
- risk assessments must be carried out
- everyone involved in work at height must be 'competent' and if being trained, must be supervised by a competent person.

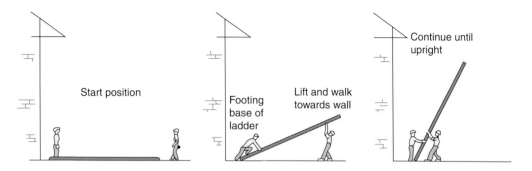

Figure 3.4 Correct procedure for erecting long or extension ladders.

Ladders

The term ladder is generally taken to include stepladders and trestles. The use of ladders for working above ground level is only acceptable for access and work of a short duration. For work over an extended period, a temporary working platform or stage is inherently a much safer means of working above ground level.

There is extensive published guidance on the safe use of ladders which is summarized below:

- It is advisable to inspect any ladder before climbing it
- The ladder should be straight and firm without signs of any damage or cracks
- All rungs and tie rods must be in place
- Ladders must not be painted because the paint may hide any defects
- Extension ladders must be erected in the closed position as shown in Fig. 3.4
- Each section of an extension ladder must overlap by at least two rungs
- The angle of the ladder to the building should be in the proportion 4 up to 1 out or 75° as shown in Fig. 3.5
- The top of the ladder must rest against a solid secure structure and not against a fragile or movable structure
- The top of the ladder must extend at least 1.05 m above the landing place or the highest rung on which the user has to stand
- Erect the ladder close to the work site and do not over-reach
- The ladder must stand on firm, level ground and be secured top and bottom
- All ladders should be tested and examined by a competent person at least yearly and the results recorded.

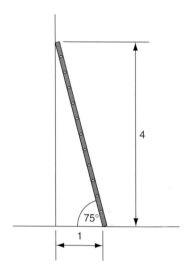

Figure 3.5 A correctly erected ladder.

> **Key fact**
>
> This is important safety information.

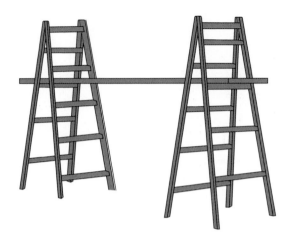

Figure 3.6 A trestle scaffold.

Stepladders

The following precautions should be observed when using stepladders:

- They should be inspected before use
- Damaged, cracked or loose jointed stepladders should not be used
- They must be extended fully
- All four legs must rest firmly and squarely on firm ground
- The stepladder should be placed at right angles to the work wherever possible
- Do not stand on the top platform unless it is designed as a working platform
- do not use the top tread, tool shelf or rear part of the steps as a foot support
- Only one person should stand on the stepladder at any one time
- The stepladder must be suitable and of an appropriate grade for the intended use.

Trestle scaffold

Two pairs of trestles or 'A' frames spanned by scaffolding boards provide a simple working platform as shown in Fig. 3.6.

- As with stepladders, they must be erected on firm level ground with the trestles fully opened
- The platform must be at least two boards or 450 mm wide
- At least one third of the trestle must be above the working platform
- The scaffold boards must be of equal length and not overhang the trestles by more than four times their thickness

Key fact

Safety First – Ladders
- The Work at Height Regulations tell us:
- ladders should only be used for access or
- for work of short duration
- 'footing' is only effective for small ladders (6 m max)
- manufactured securing devices should always be considered.

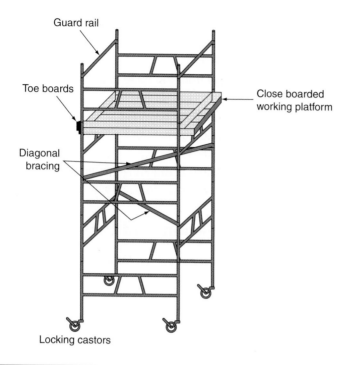

Guard rail

Toe boards

Close boarded
working platform

Diagonal
bracing

Locking castors

Figure 3.7 A mobile scaffold tower.

- The maximum span of the scaffold boards between the trestles depends upon the thickness of the boards. One metre for 32 mm boards, 1.5 m for 38 mm boards and 2.5 m for 50 mm boards.

Mobile scaffold towers

Mobile scaffold towers are normally made from light aluminium tube, slotting sections together until the required height is reached. Mobile towers are fitted with four lockable wheels; static towers have flat plates instead of wheels. A mobile scaffold tower is shown in Fig. 3.7.

This is the preferred method of working above ground for extended periods. If accidents occur it is mainly as a result of poor standards of erection or misuse. Consider the following good practice:

- The person erecting the tower must be 'competent'
- Use the tower only on level, firm ground
- If the working platform is 2 m above ground it must be close boarded and fitted with guard rails and toe boards
- The taller the tower, the more likely it is to become unstable. Outriggers can increase stability by effectively increasing the base area. Always keep within the manufacturer's safe working limits
- There must be a safe method of getting to and from the working platform. This is usually a built-in ladder which is climbed on inside the tower

- Wheel brakes must be on when the tower is in use
- Do not move the tower while it is occupied by people or there is material on the upper platform
- Push or pull the tower only from the base and look out for overhead obstructions
- Never extend the working platform with ladders or stepladders
- Ladders must not be leaned against the scaffold tower because this might push the tower over
- Ensure that the tower scaffold is regularly inspected and maintained by a trained and competent person.

Safe electrical isolation and lock off

Key fact

Safe Isolation
- never work 'live'
- isolate first
- secure the isolation
- prove the supply 'dead' before starting work.

As an electrician working on electrical equipment you must always make sure that the equipment or circuit is electrically dead before commencing work to avoid receiving an electric shock and because:

- the Electricity at Work Regulations 1989 tell us that before work commences on electrical equipment it must be disconnected from the source of supply and that disconnection must be secure. A small padlock or the removal of the fuse or MCB will ensure the security of the disconnection
- the IEE Regulations (132.15) tell us that every circuit must be provided with a means of isolation
- larger pieces of equipment and electrical machines will often have an isolator switch close by which may be locked off
- to deter anyone from trying to reconnect the supply while work is being carried out, a sign 'Danger – Electrician at Work' should be displayed on the isolator or source of the supply in addition to the small padlock
- where a test instrument or voltage indicator such as that shown in Fig. 3.8 is used to prove the supply dead, the same device must be tested to prove it is still working by using a 'proving unit' such as that shown in Fig. 3.9
- the test leads and probes of the test instrument must comply with the Health & Safety Executive Guidance Note 38 giving adequate protection to the user as shown in Fig. 3.10
- a suitable safe electrical isolation procedure is shown in Fig. 3.11.

Try this

Safe Isolation

Follow each stage of Fig. 3.11 carefully and then you should practise this safe isolation procedure at college under the guidance of your lecturer and at work under the guidance of your supervisor. It is an important safety procedure which you must learn.

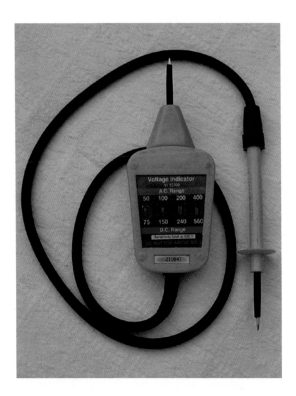

Figure 3.8 Typical voltage indicator.

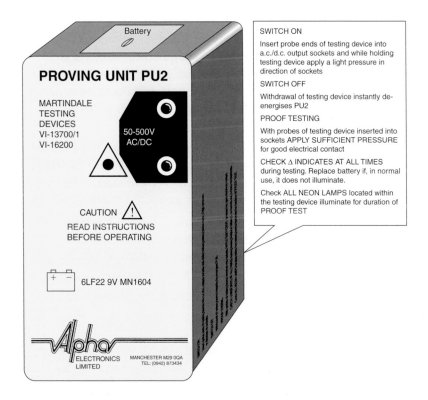

Figure 3.9 Voltage proving unit.

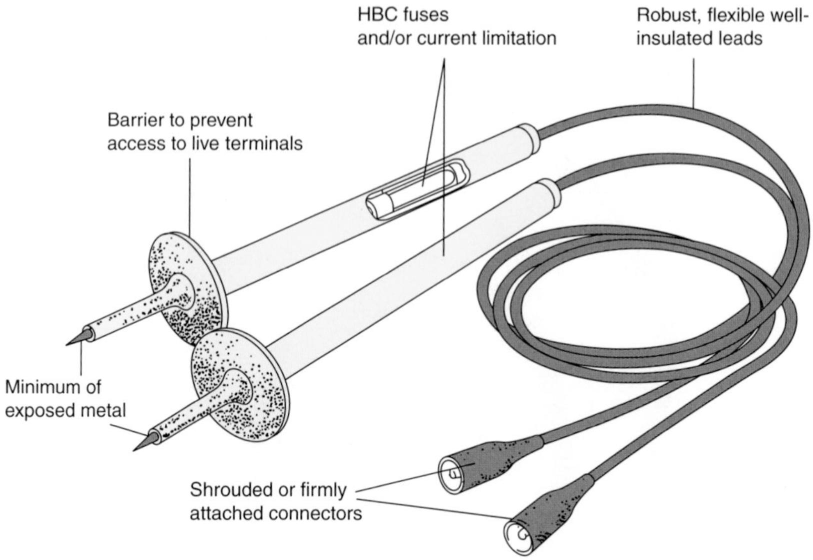

HBC fuses
and/or current limitation

Robust, flexible well-
insulated leads

Barrier to prevent
access to live terminals

Minimum of
exposed metal

Shrouded or firmly
attached connectors

Figure 3.10 GS38 recommended test probes and leads.

Electrical installation principles

A.C. theory

Commercial quantities of electricity for industry, commerce and domestic use are generated as A.C. in large power stations and distributed around the UK on the national grid to the end user. D.C. electricity has many applications where portability or an emergency stand-by supply is important but for large quantities of power it has to be an A.C. supply.

Rotating a simple loop of wire or coils of wire between the poles of a magnet such as that shown simplified in Fig. 3.12 will cut the north south lines of magnetic flux and induce an A.C. voltage in the loop or coils of wire as shown by the display on a cathode ray oscilloscope.

This is an A.C. supply, an alternating current supply. The basic principle of the A.C. supply generated in a power station is exactly the same as Fig. 3.12 except that powerful electromagnets are used and the power for rotation comes from a steam turbine.

Let us now look at some of the terms used in A.C. theory.

Phasor diagram

A phasor diagram or phasor is a straight line, having definite length and direction which represents to scale the voltage and current in an A.C. circuit.

> **Key fact**
>
> You should memorize the A.C. theory which I will cover in the next section.

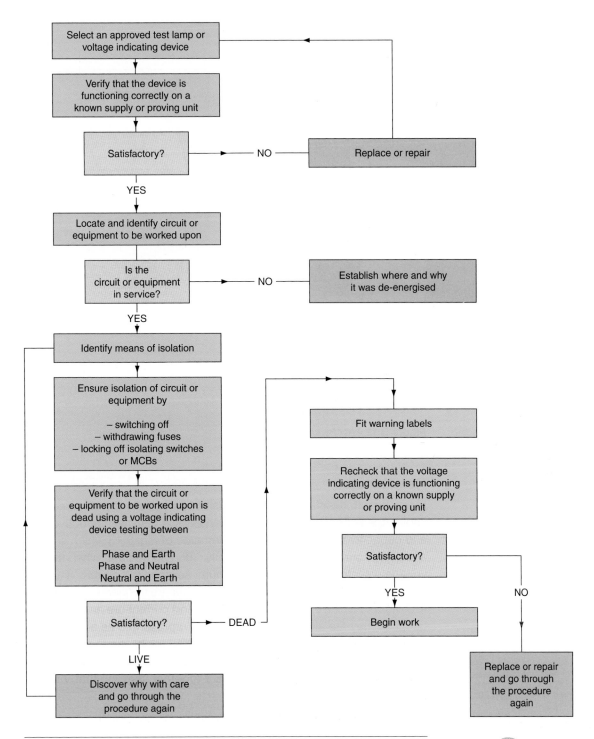

Figure 3.11 Safe electrical isolation procedure.

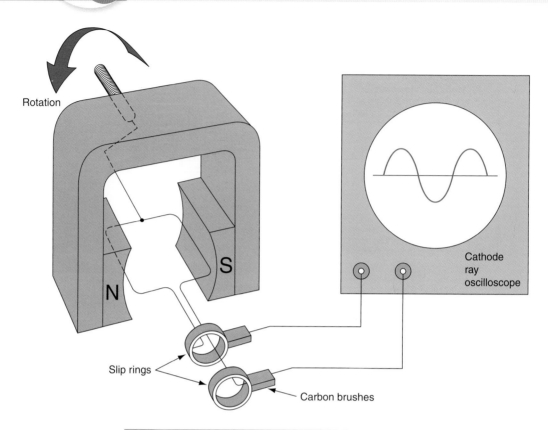

Figure 3.12 Simple A.C. generator or alternator.

Resistance

In any circuit, resistance is the opposition to current flow. Figure 3.13 shows the voltage and current relationships in resistive, inductive and capacitive circuits. Look at the left side of Fig. 3.13 which shows a resistor connected to an A.C. supply. You can see that when the voltage waveform reaches its maximum, so does the current. This always happens when resistive components are connected to an A.C. supply and we say that the voltage and current are 'in phase' because they are always together. This is represented as a phasor diagram by the bottom left-hand sketch. You might like to think of the phasors as the minute and hour hands of a clock with rotation anti-clockwise. In this case the phasors are together showing that V and I are in phase.

Water heaters, electric fires and filament lamps are resistive circuits.

Inductance

Any coil of wire possesses inductance and so we say that the opposition to current flow in an inductive circuit is called '*inductive reactance*' symbol X_L measured in ohms.

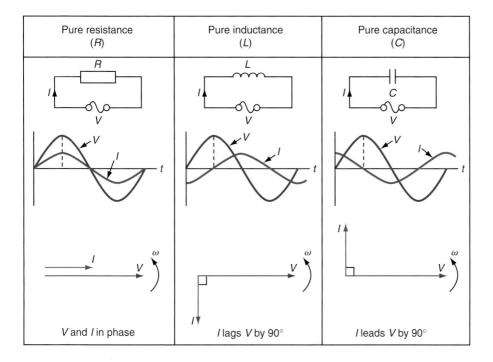

Figure 3.13 Voltage and current relationship in resistive, inductive and capacitive circuits.

When a current flows in a coil, it sets up its own voltage around the conductor which opposes the applied voltage. This causes the current to fall behind or 'lag' the applied voltage. You can see this in Fig. 3.13. Time is measured from left to right and so the current reaches its maximum value later than the voltage waveform. In fact, 90° later and so we say that in an inductive circuit, the current lags the voltage by 90°. This is represented on the phasor diagram as shown by the bottom centre sketch.

Inductive circuits are those which contain windings or coils such as electric motors, transformers or the choke of a discharge luminaire.

Capacitance

A capacitor is a component which stores an electric charge if a voltage is applied across it and so we say that the opposition to current flow in a capacitor circuit is called '**capacitive reactance**' symbol X_C measured in ohms.

When a capacitor is connected to the A.C. supply, it is continuously storing charge and then discharging as the supply moves through its positive to negative cycle. This causes the current to spring forward or to 'lead' the applied voltage. You can see this effect in Fig. 3.13. In fact, the current leads the applied voltage by 90°, just the opposite effect to an inductive circuit which is why this is important in A.C. circuit theory.

Capacitors are usually constructed from long strips of metal foil, like baking foil, insulated and then rolled up into a cylinder. You can see them in fluorescent fittings, discharge luminaires and sometimes fixed to an electric motor. The leading effect of the capacitor can be used to neutralize the lagging effect of inductors.

Power factor

Power factor or pf is defined as the cosine of the phase angle between the current and voltage. If the current lags the voltage, as can be seen in the inductive circuit of Fig. 3.13, we say that the pf is lagging and if the current leads the voltage, we say the pf is leading. The ideal situation is when the pf is neither lagging nor leading but is in phase. In this situation the pf is equal to 1.

To correct or put right the bad (lagging) power factor of an inductive circuit such as an electric motor or fluorescent light fitting, we would connect a capacitor (having a leading power factor) across the load. The leading pf of the capacitor neutralizes the lagging pf of the inductive circuit bringing the overall pf of the circuit up to, or nearly up to, 1. This is called power factor correction and 0.9, 0.95 or 1 are all acceptable values for a commercial, industrial or domestic supply.

Figure 3.14(a) shows the phasor diagram of an industrial load with a bad power factor. If a capacitor is connected in parallel with the load, the capacitor current I_c will lead the voltage by 90°. When the capacitor current is added to the load current, as shown in Fig. 3.14(b), the resultant load current has a much improved power factor. Using a slightly bigger capacitor, the load current could be pushed up until it was 'in phase' with the voltage as shown in Fig. 3.14(c).

Self and mutual inductance

If a coil of wire is wound on to an iron core, as shown in Fig. 3.15, a magnetic field will become established in the core when a current flows in the coil due to the switch being closed.

When the switch is opened, the current stops flowing and, therefore, the magnetic flux collapses. The collapsing magnetic flux induces an emf into the coil and this voltage appears across the switch contacts.

If you switch off a circuit containing fluorescent light fittings you can sometimes hear the discharge across the switch contacts (each fluorescent fitting contains a choke). The effect is known as *self-inductance*, or just *inductance*, and is the property of any coil.

When two separate coils are placed close together – as they are in a transformer – a current in one coil produces a magnetic flux which

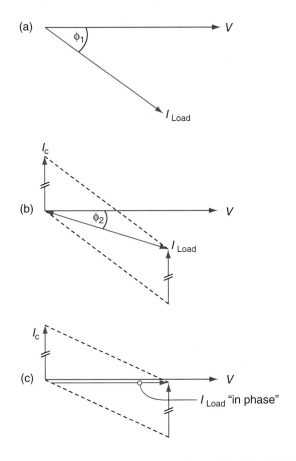

Figure 3.14 Power factor improvement using capacitors.

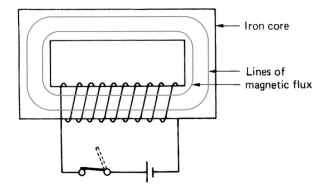

Figure 3.15 An inductive coil or choke.

links with the second coil. This induces a voltage in the second coil, and is the basic principle of the transformer action which is described later in this chapter. The two coils in this case are said to possess *mutual inductance*, as shown in Fig. 3.16.

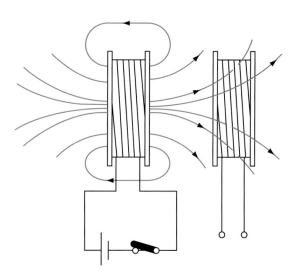

Figure 3.16 Mutual inductance between two coils.

The emf induced in a coil such as that shown on the right-hand side in Fig. 3.16 is dependent upon the rate of change of magnetic flux and the number of turns on the coil. This principle finds an application in electric motors and transformers which we will discuss next.

Electrical machines – basic operating principles

Fluorescent luminaires

A luminaire is equipment which supports an electric lamp and distributes or filters the light created by the lamp. It is essentially the 'light fitting'.

A lamp is a device for converting electrical energy into light energy. There are many types of lamps. General lighting service (GLS) lamps and tungsten halogen lamps use a very hot wire filament to create the light and so they also become very hot in use. Fluorescent tubes operate on the 'discharge' principle; that is, the excitation of a gas within a glass tube. They are cooler in operation and very efficient in converting electricity into light. They form the basic principle of most energy efficient lamps.

A fluorescent lamp is a linear arc tube, internally coated with a fluorescent powder, containing low pressure mercury vapour and argon gas. The lamp construction is shown in Fig. 3.17.

Passing a current through the electrodes of the tube produces a cloud of electrons that ionize the mercury vapour and the argon in the tube, producing invisible ultraviolet light and some blue light. The

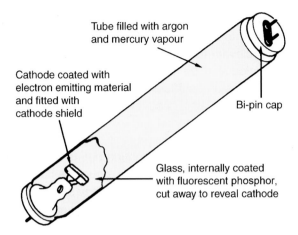

Tube filled with argon
and mercury vapour

Cathode coated with
electron emitting material
and fitted with
cathode shield

Bi-pin cap

Glass, internally coated
with fluorescent phosphor,
cut away to reveal cathode

The arc radiates much more UV than visible light:
almost all the visible light from a fluorescent tube
comes from the phosphors

Figure 3.17 Fluorescent lamp construction.

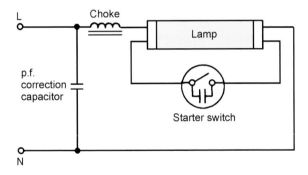

Figure 3.18 Fluorescent lamp circuit arrangements.

fluorescent powder on the inside of the glass tube is very sensitive to ultraviolet rays and converts this radiation into visible light.

Fluorescent luminaires require a simple electrical circuit to initiate the ionization of the gas in the tube and a device to control the current once the arc is struck and the lamp is illuminated. Such a circuit is shown in Fig. 3.18.

A typical application for a fluorescent luminaire is in suspended ceiling lighting modules used in many commercial buildings.

Building regulations for energy efficient lamps

Part P of the Building Regulations relates to Electrical Safety in Dwellings. All new installations must comply with the Part P

Regulations and any other relevant parts of the Building Regulations. Approved Documents L1A and L1B, Conservation of Fuel and Power 2006 are relevant to us as electricians because they say that reasonable provision shall be made to provide lighting systems with energy efficient lamps and sufficient controls so that electrical energy can be used efficiently. Part L describes methods of compliance with these regulations for both internal and external lighting. It says:

- A reasonable number of internal lighting points should be wired that will only take energy efficient lamps such as fluorescent tubes and compact fluorescent lamps (CFLs). They should be installed in the areas most frequently used and number at least one per $25\,m^2$ of dwelling area or one per four fixed luminaires.
- External lighting fixed to the building should provide reasonable provision for energy efficient lamps such as fluorescent tubes and compact fluorescent lamps (CFLs). These lamps should automatically extinguish in daylight and when not required at night, by being controlled by PIR detectors and functional switching.

The traditional light bulb, called a GLS (general lighting service) lamp, is hopelessly bad in energy efficiency terms, producing only 14 lumens of light output for every electrical watt input. Fluorescent tubes and CFLs produce more than 40 lumens of light output for every electrical watt input. The government has calculated that if every British household was to replace three 60 watt or 100 watt GLS lamps with CFLs, the energy saving would be greater than the power used by the entire street lighting network.

Mr Hilary Benn, the then Environment Secretary, announced in the spring of 2008 that the traditional GLS light bulbs of 150 watts, 100 watts, 60 watts and 40 watts will begin to be phased out by 2010. They will not be available in the shops, and so households will be forced to use energy efficient lamps in the future.

The electrical relay

A relay is an electromagnetic switch operated by a solenoid. We looked at the action of a solenoid in the last chapter at Fig. 2.12. The solenoid in a relay operates a number of switch contacts as it moves under the electromagnetic forces. Relays can be used to switch circuits on or off at a distance remotely. The energizing circuit, the solenoid, is completely separate to the switch contacts and, therefore, the relay can switch high voltage, high power circuits, from a low voltage switching circuit. This gives the relay many applications in motor control circuits, electronics and instrumentation systems. Figure 3.19 shows a simple relay.

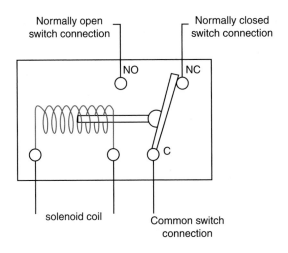

Figure 3.19 A simple relay.

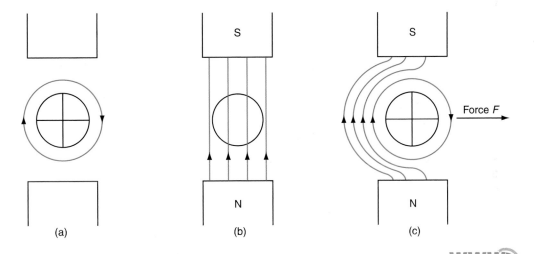

Figure 3.20 Force on a conductor in a magnetic field.

D.C. Motors

All electric motors work on the basic principle that when a current carrying conductor is placed in a magnetic field it will experience a force. An electric motor uses this magnetic force to turn the shaft of the electric motor. Let us try to understand this action. Figure 3.20(a) shows the magnetic field set up around a current carrying conductor shown in cross-section. Figure 3.20(b) shows the magnetic field due to a permanent magnet in which is placed the conductor carrying no current. Figure 3.20(c) shows the effect of the combined magnetic fields which have become distorted and, because lines of magnetic flux never cross and behave like stretched elastic bands, a force F is

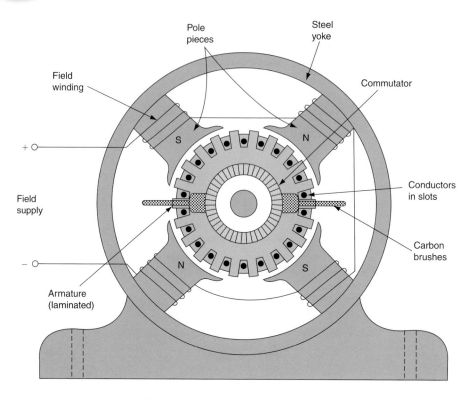

Figure 3.21 Showing D.C. machine construction.

exerted on the conductor. This is the force which turns the shaft on the electric motor.

A D.C. motor has a field winding wound on to the body or yoke of the motor and an armature winding which rotates and turns the motor shaft. Feeding the current into the armature, so that the magnetic field can be established, is the commutator and carbon brushes as shown in Fig. 3.21. D.C. motors are classified by the way in which the field and armature windings are connected. Figure 3.22 shows the connections for a series motor. Because the windings are in series, a D.C. motor will also work satisfactorily on an A.C. supply. Small D.C. series motors are also called universal motors and are used for vacuum cleaners and hand drills because they have a high starting torque for a small motor, but if the load is increased, the speed reduces.

Try this

The next time you are using a 110 V electric drill to drill a wall, switch off the hammer action and listen to the sound of the drill. As you put pressure on the drill the speed will reduce because you are loading up the motor. Reduce the pressure and the drill will speed up because it is a series motor.

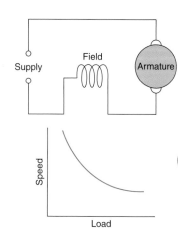

Figure 3.22 Series motor connections and characteristics.

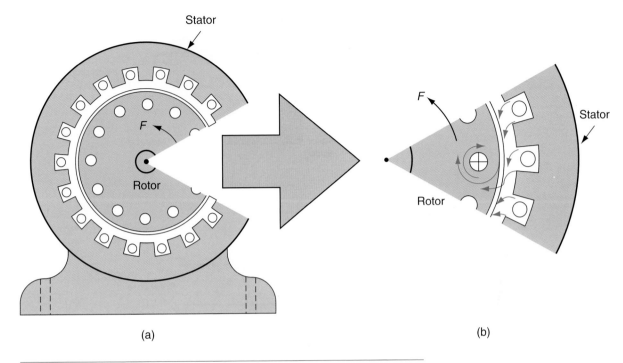

Figure 3.23 Segment taken out of induction motor.

A.C. Motors

A.C. motors are also called induction motors because of their basic principle of operation. The A.C. supply is connected to the stator windings of the motor. These are the stationary windings of the motor, like the field winding of a D.C. motor. The A.C. supply sets up a rotating magnetic field in the stator, which causes the rotor to turn. Figure 3.23 shows the magnetic flux in the stator and rotor creating the turning force or torque to drive the motor.

A D.C. motor always has a commutator and carbon brushes to connect the supply to the rotating part of the motor. These require maintenance and repair. An A.C. motor has no such equipment because the current is 'induced' into the rotor by magnetic induction, a principle discovered by Michael Faraday. No carbon brushes or commutator is a great advantage in an A.C. machine and also the construction of the rotor makes an A.C. machine very robust.

Larger motors used in industry are connected to a three phase A.C. supply, while smaller motors are connected to a single phase A.C. supply.

A.C. motors have a relatively low starting torque and are used for constant speed applications from industrial motors to air extraction

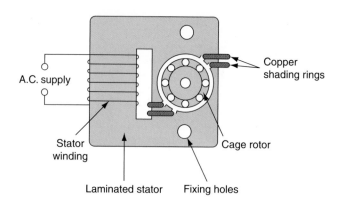

Figure 3.24 Shaded pole A.C. motor.

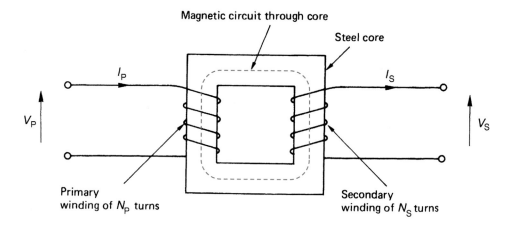

Figure 3.25 A simple transformer.

fans, fan heaters, central heating pumps, refrigerators and washing machines. Very small A.C. motors of less than 50 watts can be found in most domestic and business machines where single phase supplies are available. Figure 3.24 shows the construction of a small A.C. motor.

Transformers

A transformer is an electrical machine which is used to change the value of an alternating voltage. They vary in size from miniature units used in electronics to huge power transformers used in power stations. A transformer will only work when an alternating voltage is connected. It will not normally work from a D.C. supply such as a battery.

A transformer, as shown in Fig. 3.25, consists of two coils, called the primary and secondary coils, or windings, which are insulated from each other and wound on to the same steel or iron core.

An alternating voltage applied to the primary winding produces an alternating current, which sets up an alternating magnetic flux throughout the core. This magnetic flux induces an emf in the secondary winding by mutual inductance, which was described earlier in this chapter under the sub-heading 'Self and mutual inductance'. Since both windings are linked by the same magnetic flux, the induced emf per turn will be the same for both windings. Therefore, the emf in both windings is proportional to the number of turns. In symbols:

$$\frac{V_p}{V_s} = \frac{N_p}{N_s}$$

where:

V_p = the primary voltage
V_s = the secondary voltage
N_p = the number of primary turns
N_s = the number of secondary turns

Moving the terms around we have a general expression for a transformer:

$$\frac{V_p}{V_s} = \frac{N_p}{N_s}$$

Example

A 230 V to 12 V emergency lighting transformer is constructed with 800 primary turns. Calculate the number of secondary turns required. Collecting the information given in the question into a usable form, we have:

$$V_p = 230\,V$$

$$V_s = 12\,V$$

$$N_p = 800$$

From the general equation:

$$\frac{V_p}{V_s} = \frac{N_p}{N_s}$$

the equation for the secondary turn is

$$N_s = \frac{N_p V_s}{V_p}$$

$$\therefore\ N_s = \frac{800 \times 12V}{230V} = 42\,\text{turns}$$

42 turns are required on the secondary winding of this transformer to give a secondary voltage of 12 V.

Follow this maths

Using the general equation for a transformer given above, follow this maths carefully, step by step, in the following example.

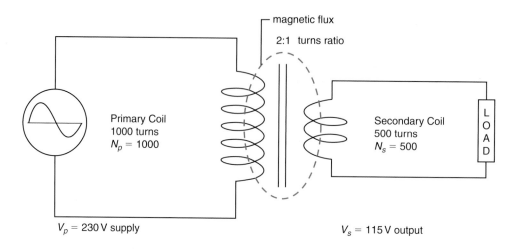

Figure 3.26 A step down transformer.

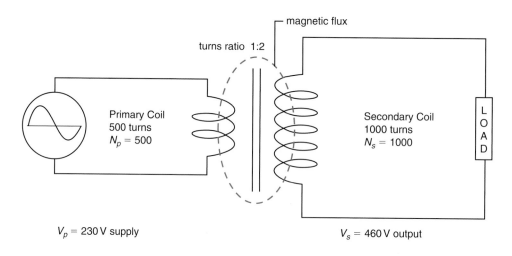

Figure 3.27 A step up transformer.

Types of transformer

Step down transformers are used to reduce the output voltage, often for safety reasons. Figure 3.26 shows a step down transformer where the primary winding has twice as many turns as the secondary winding. The turns ratio is 2:1 and, therefore, the secondary voltage is halved.

Step up transformers are used to increase the output voltage. The electricity generated in a power station is stepped up for distribution on the national grid network. Figure 3.27 shows a step up transformer where the primary winding has only half the number of turns as the secondary winding. The turns ratio is 1:2 and, therefore, the secondary voltage is doubled.

Instrument transformers are used in industry and commerce so that large currents and voltage can be measured by small electrical instruments.

A current transformer (or CT) has the large load currents connected to the primary winding of the transformer and the ammeter connected to the secondary winding. The ammeter is calibrated to take account of the turns ratio of the transformer, so that the ammeter displays the actual current being taken by the load when the ammeter is actually only taking a small proportion of the load current.

A voltage transformer (or VT) has the main supply voltage connected to the primary winding of the transformer and the voltmeter connected to the secondary winding. The voltmeter is calibrated to take account of the turns ratio of the transformer, so that the voltmeter displays the actual supply voltage.

Separated extra-low voltage (SELV) transformers If the primary winding and the secondary winding of a double wound transformer have a separate connection to earth, then the output of the transformer is effectively isolated from the input since the only connection between the primary and secondary windings is the magnetic flux in the transformer core. Such a transformer would give a very safe electrical supply which might be suitable for bathroom equipment such as shaver sockets and for construction site 110 V tools, providing that all other considerations are satisfied, such as water ingress, humidity, IP protection and robust construction.

Generation, transmission and distribution of electricity

Generation

Figure 3.12 earlier in this chapter shows a simple A.C. generator or alternator producing an A.C. waveform. We generate electricity in large modern power stations using the same basic principle of operation. However, in place of a single loop of wire, the power station alternator has a three phase winding and powerful electromagnets. The prime mover is not, of course, a simple crank handle, but a steam turbine. Hot water is heated until it becomes superheated steam, which drives the vanes of a steam turbine which is connected to the alternator. The heat required to produce the steam may come from burning coal or oil or from a nuclear reactor. Whatever the primary source of energy is, it is only being used to drive a turbine which is connected to an alternator, to generate electricity.

Figure 3.28 Transmission line steel pylon.

Try this

Electricity Distribution

- *why do you think supplies to remote farms are distributed overhead when electricity distribution in villages, towns and cities is underground?*
- *where is the local distribution sub-station of your college? Ask your lecturer.*
- *where is the local distribution sub-station near where you live? They are never obvious and often hidden away*
- *you could make notes in the margin here if that helps.*

Transmission

Electricity is generated in the power station alternator at 25 kV. This electrical energy is fed into a transformer to be stepped up to a very high voltage for transmission on the national grid network at 400 kV, 275 kV or 132 kV. These very high voltages are necessary because, for a given power, the current is greatly reduced, which means smaller grid conductors and the transmission losses are reduced.

The national grid network consists of over 5000 miles of overhead aluminium conductors suspended from steel pylons which link together all the power stations. Figure 3.28 shows a transmission line steel pylon.

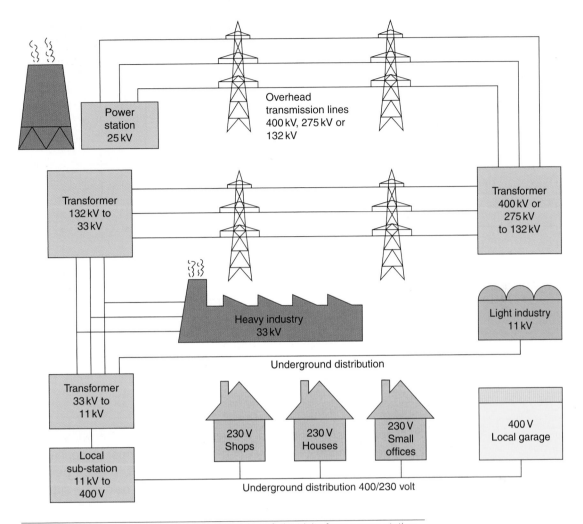

Figure 3.29 Simplified diagram of the distribution of electricity from power station to consumer.

Electricity is taken from the national grid by appropriately located sub-stations which eventually transform the voltage down to 11 kV at a local sub-station. At the local sub-station the neutral conductor is formed for single phase domestic supplies and three phase supplies to shops, offices and garages. These supplies are usually underground radial supplies from the local sub-station but in rural areas we still see transformers and overhead lines suspended on wooden poles. Figure 3.29 gives an overview of the system from power station to consumer.

Distribution to the consumer

The electricity leaves the local sub-station and arrives at the consumer's mains intake position. The final connections are usually

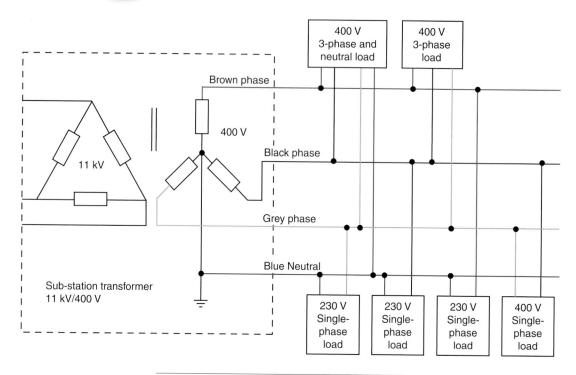

Figure 3.30 Three phase four wire distribution.

by simple underground radial feeders at 400 V/230 V. The 400 V/230 V is derived from the 11 kV/400 V sub-station transformer by connecting the secondary winding in star as shown in Fig. 3.30. The star point is earthed to an earth electrode sunk into the ground below the sub-station and from this point is taken the fourth conductor, the neutral. Loads connected between phases are fed at 400 V and those fed between one phase and neutral at 230 V. A three phase 400 V supply is used for supplying small industrial and commercial loads such as garages, schools and blocks of flats. A single phase 230 V supply is usually provided for individual domestic consumers.

At the mains intake position the supplier will provide a sealed HBC fuse and a sealed energy meter to measure the consumer's electricity consumption. It is after this point that we reach the consumer's installation.

Balancing single phase loads

A three phase load such as a motor has equally balanced phases since the resistance of each phase winding will be the same. Therefore, the current taken by each phase will be equal. When connecting single phase loads to a three phase supply, care should be taken to distribute the single phase loads equally across the three

Figure 3.31 The provision of a safe electrical installation is fundamental to the whole concept of using electricity.

phases so that each phase carries approximately the same current. Equally distributing the single phase loads across the three phase supply is known as 'balancing' the load. A lighting load of 18 luminaires would be 'balanced' if six luminaires were connected to each of the three phases.

Protecting electrical equipment, circuits and people

The provision of a safe electrical system is fundamental to the whole concept of using electricity in and around buildings safely. The electrical installation as a whole must be protected against overload and short circuit damage and the people using the installation must be protected against electric shock. An installation which meets the requirements of the IEE Regulations, Requirements for Electrical Installations, will be so protected. The method most universally used in the UK to provide for the safe use of electrical energy is Basic Protection and Fault Protection. So let us look at these two essential safety elements.

Shock Protection: is protection from an electric shock and is provided by both Basic Protection and Fault Protection.

Basic Protection: is provided by insulating all 'live' parts in accordance with Section 416 of the IEE Regulations.

Fault Protection: is provided by protective equipotential bonding and the automatic disconnection of the supply by a fuse, circuit breaker or residual current device in accordance with IEE Regulations 411.3 to 6. Figures 4.5, 4.6 and 4.7 in Chapter 4 show the method of connecting protective equipotential bonding conductors to typical domestic installations.

Earthing and bonding

Chapter 54 of the IEE Regulations describes the earthing arrangements for an electrical installation. Let us define some of the terms used in earthing and bonding.

Earth: The general mass of the planet earth is considered to be a large conductor at zero potential (potential means voltage in this case). The act of earthing connects together all metalwork, other than that intended to carry current, to the general mass of earth so that a dangerous potential difference (voltage) cannot exist between different metal parts, or between metal parts and earth.

Earthing: The IEE Regulations define earthing as the act of connecting the exposed conductive parts of an installation to the main earthing terminal of the installation.

Exposed Conductive Parts: The IEE Regulations define these as a conductive part which may be touched and which is not live under normal conditions, but may become live under fault conditions. This means the metalwork of an appliance or the metalwork of the electrical installation such as the conduit, trunking or metal boxes of the electrical system, all of which must be connected to the main earthing terminal of the installation.

Circuit Protective Conductor (CPC): This is a protective conductor connecting exposed conductive parts to the main earthing terminal. It will be a green and yellow insulated conductor of appropriate size within the cable.

Extraneous Conductive Parts: This is the structural steelwork of a building and other service pipes used for gas, water, etc. (radiators and sinks). They do not form a part of the electrical installation, but may introduce a potential to the electrical installation. To eliminate this hazard we provide protective equipotential bonding.

Protective Equipotential Bonding: This is equipotential bonding for the purpose of safety. It is an electrical connection which maintains exposed conductive parts and extraneous conductive parts at the same potential. To do this, we connect a green and yellow insulated cable of appropriate size, but probably $10\,mm^2$, to all extraneous and exposed conductive parts and connect this to the main earthing terminal of the installation. By connecting to earth all metalwork not intended to carry current, a safe path is provided for any leakage currents which can be detected and disconnected by fuses, circuit breakers and RCDs.

Bonding Conductor: This is the protective conductor providing equipotential bonding.

A good earth path, that is a low resistance earth path, will allow high fault currents to flow, which will cause protective devices to operate quickly and remove the potential hazard quickly.

Bonding safety and other trades

- The application of protective equipotential bonding to earth is one of the important principles of fault protection.
- Protective equipotential bonding to earth is equipotential bonding for the purpose of safety (IEE Regulation 131.2.2).
- Main protective bonding conductors shall connect the main earthing terminal of the electrical installation to the water installation, the gas installation, other pipework and ducting, central heating systems and exposed metallic structural parts (IEE Regulation 411.3.1.2).
- If other trades find that they must remove the main equipotential bonding conductors in order to repair, let us say, the gas or water services, then ONE of the following TWO actions must be taken:

1   the electrical supply to the whole electrical installation must be switched off while the non-electrical work is carried out, OR

2.1   appropriate temporary equipotential bonding conductors must be installed by a competent person to maintain the integrity of the electrical installation before the existing bonding conductors are removed.

2.2   upon completion of the non-electrical repair work, the main equipotential bonding conductors must be restored to their original position using approved and secure bonding clamps to cleaned pipework in accordance with IEE Regulation 514.13.1.

2.3   the temporary bonding conductors may then be removed.

Key fact

Bonding Safety and Other Trades
- main protective bonding conductors must not be removed unless:
- the electrical supply to the whole installation is first switched off, OR
- temporary $10\,mm^2$ bonding conductors and robust clamps are first installed.

Overcurrent protection

All circuit conductors must be protected against overcurrent, that is a current exceeding the rated value (IEE Regulation 430.3). Fuses and circuit breakers provide overcurrent protection when situated in the live conductor. They must not be connected in the neutral conductor.

Overcurrent conditions arise because of an **overload** or a **short circuit** in the electrical circuit.

By definition an overload current occurs in a circuit which is carrying more current than it was designed to carry. The excess current may be a result of too many pieces of equipment being connected to the circuit or because a piece of equipment has become faulty. An overload current will result in currents of two or three times the rated current flowing. This will cause the cable temperature to rise, leading to an increased risk of fire.

By definition a short circuit current occurs in a circuit as a result of a fault or damage to the circuit which could not have been predicted before the event. The short circuit current may be the result of a nail being driven through an energized cable, making contact with the live conductor and either the neutral or earth conductors. A short circuit current will result in currents hundreds of times greater than the rated current flowing. To avoid the risk of fire or electric shock, these overcurrents must be interrupted quickly and the circuit made dead.

Try this

Try writing out these definitions – just the essential bits to help you to remember them – for overcurrent, overload, short circuit current

Devices which provide overcurrent protection are:

- Semi-enclosed fuses to BS 3036
- Cartridge fuses to BS 88-3
- MCBs (miniature circuit breakers) to BS EN 60898

By definition a fuse is the weakest link in the circuit. Under fault conditions it will melt when an overcurrent flows, protecting the circuit conductors from damage.

Semi-enclosed fuses (BS 3036)

The semi-enclosed fuse consists of a fuse wire, called the fuse element, secured between two screw terminals in a fuse carrier. The fuse element is connected in series with the load and the thickness of the element is sufficient to carry the normal rated circuit current.

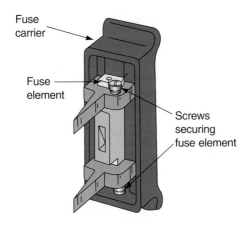

Figure 3.32 A semi-enclosed fuse.

When a fault occurs an overcurrent flows and the fuse element becomes hot and melts or 'blows'.

This type of fuse is illustrated in Fig. 3.32. The fuse element should consist of a single strand of plain or tinned copper wire having a diameter appropriate to the current rating of the fuse. This type of fuse was very popular in domestic installations, but less so these days because of their disadvantages.

Advantages of semi-enclosed fuses:

- They are very cheap compared with other protective devices both to install and to replace
- There are no mechanical moving parts
- It is easy to identify a 'blown' fuse

Disadvantages of semi-enclosed fuses:

- The fuse element may be replaced with wire of the wrong size either deliberately or by accident
- The fuse element weakens with age due to oxidization, which may result in a failure under normal operating conditions
- The circuit cannot be restored quickly since the fuse element requires screw fixing
- They have low breaking capacity since, in the event of a severe fault, the fault current may vaporize the fuse element and continue to flow in the form of an arc across the fuse terminals
- They are not guaranteed to operate until up to twice the rated current is flowing
- There is a danger from scattering hot metal if the fuse carrier is inserted into the base when the circuit is faulty.

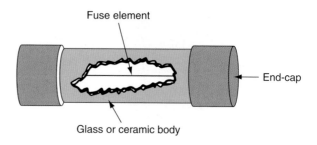

Figure 3.33 Cartridge fuse.

Try this

Definitions

- *write out the definition of a fuse*
- *and an MCB.*
- *That is, what is a fuse and an MCB?*
- *Perhaps you could do this in the margin here.*

Cartridge fuses (BS 88-3)

The cartridge fuse breaks a faulty circuit in the same way as a semi-enclosed fuse, but its construction eliminates some of the disadvantages experienced with an open-fuse element. The fuse element is encased in a glass or ceramic tube and secured to end-caps which are firmly attached to the body of the fuse so that they do not blow off when the fuse operates. Cartridge fuse construction is illustrated in Fig. 3.33. With larger size cartridge fuses, lugs or tags are sometimes brazed on the end-caps to fix the fuse cartridge mechanically to the carrier. They may also be filled with quartz sand to absorb and extinguish the energy of the arc when the cartridge is brought into operation.

Advantages of cartridge fuses:

- They have no mechanical moving parts
- The declared rating is accurate
- The element does not weaken with age
- They have small physical size and no external arcing which permits their use in plug tops and small fuse carriers
- Their operation is more rapid than semi-enclosed fuses. Operating time is inversely proportional to the fault current
- They are easy to replace

Disadvantages of cartridge fuses:

- They are more expensive to replace than fuse elements that can be re-wired

Figure 3.34 MCBs – B breaker, fits Wylex standard consumer unit. Courtesy of Wylex.

- They can be replaced with an incorrect cartridge
- The cartridge may be shorted out by wire or silver foil in extreme cases of bad practice
- It is not possible to see if the fuse element is broken.

Miniature circuit breakers BS EN 60898

The disadvantage of all fuses is that when they have operated they must be replaced. An MCB overcomes this problem since it is an automatic switch which opens in the event of an excessive current flowing in the circuit and can be closed when the circuit returns to normal.

An MCB of the type shown in Fig. 3.34 incorporates a thermal and magnetic tripping device. The load current flows through the thermal and the electromagnetic devices in normal operation but under overcurrent conditions they activate and trip the MCB.

The circuit can be restored when the fault is removed by pressing the ON toggle. This latches the various mechanisms within the MCB and 'makes' the switch contact. The toggle switch can also be used to disconnect the circuit for maintenance or isolation or to test the MCB for satisfactory operation.

Advantages of MCBs:

- They have factory set operating characteristics
- Tripping characteristics and therefore circuit protection is set by the installer

- The circuit protection is difficult to interfere with
- The circuit is provided with discrimination
- A faulty circuit may be quickly identified
- A faulty circuit may be easily and quickly restored
- The supply may be safely restored by an unskilled operator

Disadvantages of MCBs:

- They are relatively expensive but look at the advantages to see why they are so popular these days
- They contain mechanical moving parts and therefore require regular testing to ensure satisfactory operation under fault conditions.

Assessment questions

Identify the statements as true or false. If only part of the statement is false, tick false

1   The most common cause of accidents in the workplace is:
    - Slips, trips and falls
    - Moving objects by hand
    - Storing equipment badly, which then becomes unstable
    - An electric shock

    True   ☐   False   ☐

2   A hazard is defined as the possibility of harm actually being done. It may be high or low but to the Health and Safety Inspector, only low will be acceptable.

    True   ☐   False   ☐

3   Risk is something that might cause harm to a worker. For example, wet floors, working above ground or using machinery

    True   ☐   False   ☐

4   Many accidents are caused by lifting and moving heavy objects manually (that is, by hand). If a job involves a lot of manual handling, workers must be:
    - Trained to move loads safely
    - Or be provided with a suitable mechanical aid such as a sack truck

    True   ☐   False   ☐

5   When working above ground, ladders must only be used for access to the workplace or for working above ground for short periods only.

    True   ☐   False   ☐

6   When erecting a ladder, it must be made secure and have an angle against the wall of 90° or 1 up to every 4 out.

    True   ☐   False   ☐

7   A safe and secure electrical isolation procedure will always include a means of 'locking off' the electrical supply.

    True   ☐   False   ☐

8 When connecting single phase loads across a three phase supply they should be equally balanced so that the currents in the three phase supply remain approximately the same.

True ☐   False ☐

9 By definition, a fuse is the weakest link in a circuit. Under fault conditions it will melt, protecting the circuit from damage.

True ☐   False ☐

10 By definition, an overload current occurs as a result of a fault, and a short circuit current occurs in a circuit which is carrying more current than it was designed to carry.

True ☐   False ☐

Multiple choice assessment questions

Tick the correct answer. Note that more than ONE answer may be correct

11 **Health and safety legislation requires all employers who employ more than five workers to:**
   a. pay each worker 10% above the national rate for the job ☐
   b. display a health and safety law poster ☐
   c. prepare a written health and safety policy statement ☐
   d. carry out risk assessments ☐

12 **Slips, trips and falls:**
   a. do not happen at work because of safety legislation ☐
   b. always happen to someone else ☐
   c. are one of the most common causes of accidents in the workplace ☐
   d. must be reported to the HSE if they result in an absence from work of more than three days ☐

13 **A 'competent' worker is one who:**
   a. cannot do the job or task ☐
   b. can do the job or task more quickly than anyone else ☐
   c. has been trained to do a job or task successfully ☐
   d. is quarrelsome and likely to cause an argument at work ☐

14 **Hazard may be defined as:**
   a. anything that can cause harm ☐
   b. the chance, large or small, of harm actually being done ☐
   c. someone who has the necessary training and expertise to safely carry out an activity ☐
   d. the rules and regulations of the working environment ☐

15 **Risk may be defined as:**
   a. anything that can cause harm ☐
   b. the chance, large or small, of harm actually being done ☐

c. someone who has the necessary training and expertise
    to safely carry out an activity ☐

d. the rules and regulations of the working environment ☐

16 Hazard risk assessment is:

a. the harm which might be done to an employee not
    wearing PPE ☐

b. the hazard created when someone lifts a very heavy object ☐

c. the process of systematically examining the workplace
    for possible dangers ☐

d. the risk of harm being done to someone in the workplace ☐

17 A positive attitude to safety at work:

a. is the duty of every employer ☐

b. is the duty of every employee ☐

c. increases accidents at work ☐

d. reduces accidents at work ☐

18 Manual handling is the process of:

a. following instructions from a reference book ☐

b. following instructions from manufacturers' data sheets ☐

c. lifting, transporting or supporting loads by hand or bodily
    force ☐

d. moving a heavy load on a sack truck or other mechanical
    aid ☐

19 The main hazards associated with working at height are:

a. ladders not being secured top and bottom ☐

b. extension ladders not being fully extended ☐

c. people falling ☐

d. objects falling on to people ☐

**20 The angle of a ladder to the building upon which it is resting
should be in the proportion of:**

a. 1 up to 4 out ☐

b. 4 up to 75 out ☐

c. 4 up to 1 out ☐

d. 75 up to 4 out ☐

**21 The angle which a correctly erected ladder should make
with level ground is:**

a. 41° ☐

b. 45° ☐

c. 57° ☐

d. 75° ☐

22 Extension ladders should be erected:

   a.   one section at a time ☐

   b.   in the closed position ☐

   c.   in the open position ☐

   d.   against a solid structure ☐

23 Ladders must extend above the landing place or highest rung on which the user will stand by:

   a.   1.00 m ☐

   b.   1.05 m ☐

   c.   4.00 m ☐

   d.   75.00 m ☐

24 All ladders, including stepladders, must be tested and inspected at least yearly by:

   a.   the managing director of the company ☐

   b.   the site supervisor or engineer ☐

   c.   any competent person ☐

   d.   the senior electrical trainee ☐

25 To use stepladders safely:

   a.   all four legs must rest firmly and squarely on firm ground ☐

   b.   always stand on the top platform ☐

   c.   always stand on the tool platform ☐

   d.   they must be fully extended ☐

26 The preferred method of working above ground level for an extended period is:

   a.   a stepladder ☐

   b.   an extension ladder ☐

   c.   a trestle scaffold ☐

   d.   a scaffold tower ☐

27 For good stability mobile towers must have a base width to tower height ratio of:

   a.   1:2 ☐

   b.   1:3 ☐

   c.   1:4 ☐

   d.   1:5 ☐

28 To verify or prove a successful electrical isolation you would use a:

   a.   voltage indicator such as that shown in Fig. 3.8 ☐

   b.   voltage proving unit such as that shown in Fig. 3.9 ☐

   c.   set of GS 38 test leads ☐

   d.   small padlock ☐

29 **To secure an electrical isolation you would use a:**
   a.   voltage indicator such as that shown in Fig. 3.8 ☐
   b.   voltage proving unit such as that shown in Fig. 3.9 ☐
   c.   set of GS 38 test leads ☐
   d.   small padlock ☐

30 **Where a test instrument or voltage indicator is used to prove a supply dead, the same device must be tested to show that it still works using a:**
   a.   voltage indicator such as that shown in Fig. 3.8 ☐
   b.   voltage proving unit such as that shown in Fig. 3.9 ☐
   c.   set of GS 38 test leads ☐
   d.   small padlock ☐

31 **To give adequate protection to the person carrying out a safe isolation procedure, the test instrument must incorporate a:**
   a.   voltage indicator such as that shown in Fig. 3.8 ☐
   b.   voltage proving unit such as that shown in Fig. 3.9 ☐
   c.   set of GS 38 test leads ☐
   d.   small padlock ☐

32 **When a resistor circuit is connected to an A.C. supply:**
   a.   an alternating current flows ☐
   b.   the current and voltage are 'in phase' ☐
   c.   the current falls behind or 'lags' the voltage ☐
   d.   the current springs forward or 'leads' the voltage ☐

33 **When an inductive coil is connected to an A.C. supply:**
   a.   an alternating current flows ☐
   b.   the current and voltage are 'in phase' ☐
   c.   the current falls behind or 'lags' the voltage ☐
   d.   the current springs forward or 'leads' the voltage ☐

34 **When a capacitor is connected to an A.C. supply:**
   a.   an alternating current flows ☐
   b.   the current and voltage are 'in phase' ☐
   c.   the current falls behind or 'lags' the voltage ☐
   d.   the current springs forward or 'leads' the voltage ☐

35 **The cosine of the phase angle between the current and voltage is one definition of:**
   a.   resistance R ☐
   b.   inductive reactance X_L ☐
   c.   capacitive reactance X_C ☐
   d.   power factor pf ☐

36 **The opposition to current flow in a coil connected to an A.C. supply is:**

   a. resistance R ☐

   b. inductive reactance X_L ☐

   c. capacitive reactance X_C ☐

   d. power factor pf ☐

37 **The opposition to current flow in a capacitive A.C. circuit is:**

   a. resistance R ☐

   b. inductive reactance X_L ☐

   c. capacitive reactance X_C ☐

   d. power factor pf ☐

38 **Power factor correction is applied to an A.C. circuit in order to create the conditions where:**

   a. current leads the voltage ☐

   b. voltage leads the current ☐

   c. voltage and current are 'out of phase' ☐

   d. voltage and current are 'in phase' ☐

39 **Power factor correction for a fluorescent luminaire is achieved by connecting across the mains supply a:**

   a. choke ☐

   b. starter canister ☐

   c. ballast ☐

   d. capacitor ☐

40 **An electromagnetic switch operating a number of electrical contacts is one definition of:**

   a. A.C. motors ☐

   b. D.C. machines ☐

   c. a relay ☐

   d. a transformer ☐

41 **When a current carrying conductor is placed in a magnetic field it will experience a force. This is the basic principle of:**

   a. A.C. motors ☐

   b. D.C. machines ☐

   c. a relay ☐

   d. a transformer ☐

42 **Connecting an A.C. supply to the stator windings of the machine induces a rotating magnetic field which causes the rotor to turn. This is the basic principle of:**

   a. A.C. motors ☐

   b. D.C. machines ☐

c   a relay ☐

d.   a transformer ☐

43   An electrical machine has two separate windings on a common iron core. An A.C. voltage in one winding induces an A.C. voltage in the other winding, which is proportional to the number of turns. This is the basic principle of:

a.   A.C. motors ☐

b.   D.C. machines ☐

c.   a relay ☐

d.   a transformer ☐

44   A step down transformer has a turns ratio of 800:42. When a 230V supply is connected the secondary voltage will be:

a.   5.47V ☐

b.   12.00V ☐

c.   19.00V ☐

d.   43.83V ☐

45   A step up transformer has a turns ratio of 1:10. If the primary voltage is 100V the secondary voltage will be:

a.   1V ☐

b.   10V ☐

c.   100V ☐

d.   1000V ☐

46   A step down transformer has a turns ratio of 10:1. If the primary voltage is 100V, the secondary voltage will be:

a.   1V ☐

b.   10V ☐

c.   100V ☐

d.   1000V ☐

47   A construction site transformer has a turns ratio of 535:256. When connected to a 230V supply the transformer will deliver a secondary voltage of:

a.   2.09V ☐

b.   110.06V ☐

c.   480.60V ☐

d.   595.50V ☐

48   An isolating transformer for a tungsten halogen dichroic reflector lamp has a turns ratio of 38:2. Calculate the secondary voltage when connected to a 230V supply:

a.   12.1V ☐

b.   19.0V ☐

c. 43.7 V ☐

d. 76.0 V ☐

49 A voltage transformer is connected to the 230 V mains supply so that it might supply an electrical measuring instrument operating at 2 V. The turns ratio of this transformer will therefore be:

a. 1:46 ☐

b. 46:1 ☐

c. 1:115 ☐

d. 115:1 ☐

50 An isolating transformer can be found:

a. in local sub-stations operating at 11 kV : 415 V ☐

b. in large commercial power stations ☐

c. in sub-stations connecting power lines to the national grid ☐

d. in a bathroom shaver unit ☐

51 When connecting single phase loads to a three phase supply, we must take care to distribute the single phase loads equally across the three phases so that each phase carries approximately the same current. This is called:

a. generation of the phase loads ☐

b. transmission of the phases ☐

c. distribution of the load ☐

d. balancing of the load ☐

52 The metal structural steelwork of a building is called:

a. the general mass of earth ☐

b. the circuit protective conductor (CPC) ☐

c. exposed conductive parts ☐

d. extraneous conductive parts ☐

53 The protective conductor connecting exposed conductive parts of equipment to the main earthing terminal is called:

a. the general mass of earth ☐

b. the circuit protective conductor (CPC) ☐

c. exposed conductive parts ☐

d. extraneous conductive parts ☐

54 The trunking and conduit of an electrical installation are called:

a. the general mass of earth ☐

b. the circuit protective conductor (CPC) ☐

c. exposed conductive parts ☐

d. extraneous conductive parts ☐

55 **The metalwork of a piece of electrical equipment is called:**

    a.   the general mass of earth ☐

    b.   the circuit protective conductor (CPC) ☐

    c.   exposed conductive parts ☐

    d.   extraneous conductive parts ☐

56 **An electrical connection which maintains exposed and extraneous conductive parts at the same potential is called:**

    a.   CPC (circuit protective conductor) ☐

    b.   earth conductors ☐

    c.   protective equipotential bonding ☐

    d.   supplementary bonding ☐

57 **An overload current may be defined as:**

    a.   a current in excess of at least 15 A ☐

    b.   a current which exceeds the rated value in an otherwise healthy circuit ☐

    c.   an overcurrent resulting from a fault between live and neutral conductors ☐

    d.   a current in excess of 60 A ☐

58 **A short circuit may be defined as:**

    a.   a current in excess of at least 15 A ☐

    b.   a current which exceeds the rated value in an otherwise healthy circuit ☐

    c.   an overcurrent resulting from a fault between live and neutral conductors ☐

    d.   a current in excess of 60 A ☐

59 **It is the weakest link in the circuit. Under fault conditions it will melt, protecting the circuit conductors from damage. This is one description of:**

    a.   an electromagnetic relay ☐

    b.   an MCB (miniature circuit breaker) ☐

    c.   a fuse ☐

    d.   an isolating switch ☐

60 **It is an automatic switch which opens when an overcurrent flows in the circuit. This is one description of:**

    a.   a fuse ☐

    b.   an MCB ☐

    c.   an isolating switch ☐

    d.   a thermostat ☐

CHAPTER 4

Installation (building and structures)

To support your learning, instructional videos and animated versions of some of the figures from this chapter are viewable for free from the book's companion site.

When you see the logo, visit the website below to access them:
www.elsevierdirect.com/
companions/9780080969404

This chapter is concerned with the underlying principles related to electrical installation work. Understanding the laws and regulations, the different types of installation and cabling and equipment used in electrical installation work.

Regulations and responsibilities

In Chapter 1 of this book we looked at a number of the regulations which control the electrotechnical industry. The Electricity at Work (EAW) Regulations are legally binding regulations which concern all aspects of electrical systems, equipment and installations, which have been, or are to be, energized.

Electricity at Work Regulations and Codes of Practice

The EAW Regulations came into force in 1 April 1990. The purpose of the regulations is to require precautions to be taken against the risk of death or personal injury from electricity in work activities.

The EAW Regulations are made under the Health & Safety at Work Act 1974 and are statutory regulations (see Chapter 1 for a description of statutory laws). In the introduction at Section 7 of the

Figure 4.1 Electrical regulations are enforced by law.

EAW Regulations it sets out the position of the IEE Regulations in the following terms:

> *The Institution of Electrical Engineers' Regulations for Electrical Installations (the IEE Wiring Regulations) are non-statutory regulations relating principally to the design, selection, erection and inspection and testing of electrical installations. The IEE Wiring Regulations are a code of practice which is widely recognised and accepted in the UK and compliance with them is likely to achieve compliance with relevant aspects of the EAW Regulations 1989.*

If a contract specifies that the work will be carried out in accordance with BS 7671, the IEE Wiring Regulations, then this would be legally binding and the IEE Wiring Regulations will then become a legal requirement of the contract.

IEE Regulations (BS 7671)

The first edition of the IEE Regulations was issued in 1882 as the Rules and Regulations for the Prevention of Fire Risks Arising from Electric

Lighting. In the intervening 125 years there have been many new editions and we are currently using the 17th Edition from 1 July 2008.

The main reason for incorporating the IEE Wiring Regulations into British Standard 7671 was to create harmonization with European Standards. The IEE Wiring Regulations (BS 7671: 2008) are compliant with European Standards. British Standards having a BS EN number are European harmonized standards and all such standards will become common throughout Europe.

The IEE Wiring Regulations (BS 7671) are the 'Electrician's Bible' and provides the authoritative framework for anyone working in the electrotechnical industry.

To assist workers in the electrotechnical industry with their understanding of the relevant regulations many guidance booklets have been published, particularly:

- The *On Site Guide* published by the IEE
- Guidance Note 1: Selection and erection of equipment
- Guidance Note 2: Isolation and Switching
- Guidance Note 3: Inspection and Testing
- Guidance Note 4: Protection against Fire
- Guidance Note 5: Protection against Electric Shock
- Guidance Note 6: Protection against Overcurrent
- Guidance Note 7: Special Locations
- Guidance Note 8: Earthing and Bonding

All the above publications are published by the IEE and are available from IEE Publications, Michael Faraday House, Six Hills Way, Stevenage, SG1 2AY. Telephone (01438) 755540 or at www.iee.org/books

The 'Electrician's Guide to Good Electrical Practice', known as a 'toolbox guide' is published by the trade union Unite.

On-site communications

Read through the 'Communications and technical information' section of Chapter 1 before going on to this new work.

Good communication is about transferring information from one person to another. How many hours or days did you spend on a particular job last week? How does your boss know how many hours of work you put in on that job, so that a charge to the customer for your time can be made? How much material did you use on that job last week? How does your boss know how much material you used, so that a charge to the customer can be made for it?

Most electrical companies have standard forms which help them to keep track of time put in and materials used. When completing standard forms, follow the instructions given and make sure that your writing is legible – print if it makes your writing clearer. Finally, read through the form to make sure that you have completed all the relevant sections. Now, let us look at five standard forms used by most electrotechnical companies.

Time sheets

A time sheet is a standard form completed by each employee to inform the employer of the actual time spent working on a particular contract or site. This helps the employer to bill the hours of work to an individual job. It is usually a weekly document and includes the number of hours worked, the name of the job and any travelling expenses claimed. Office personnel require time sheets such as that shown in Fig. 4.2 so that wages can be made up.

Job sheets

A job sheet or job card such as that shown in Fig. 4.3 carries information about a job which needs to be done, usually a small job. It gives the name and address of the customer, contact telephone numbers, often a job reference number and a brief description of the work to be carried out. A typical job sheet work description might be:

- Job 1 – Upstairs lights not working
- Job 2 – Funny fishy smell from kettle socket in kitchen

The time spent on each job and the materials used are sometimes recorded on the job sheets, but alternatively, a daywork sheet can be used. This will depend upon what is normal practice for the particular electrical company. This information can then be used to 'bill' the customer for work carried out.

Daywork sheets

Daywork is one way of recording variations to a contract; that is, work done which is outside the scope of the original contract. If daywork is to be carried out, the site supervisor must first obtain a signature from the client's representative, for example the architect, to authorize the extra work. A careful record must then be kept on the daywork sheets of all extra time and materials used so that the client can be billed for the extra work and materials. A typical daywork sheet is shown in Fig. 4.4.

Delivery notes

When materials are delivered to site, the person receiving the goods is required to sign the driver's 'Delivery Note'. This record is used to

TIME SHEET

FLASH-BANG ELECTRICAL

Employee's name (Print) _____

Week ending _____

Day	Job number and/or Address	Start time	Finish time	Total hours	Travel time	Expenses
Monday						
Tuesday						
Wednesday						
Thursday						
Friday						
Saturday						
Sunday						

Employee's signature _____ Date _____

Figure 4.2 Time sheet.

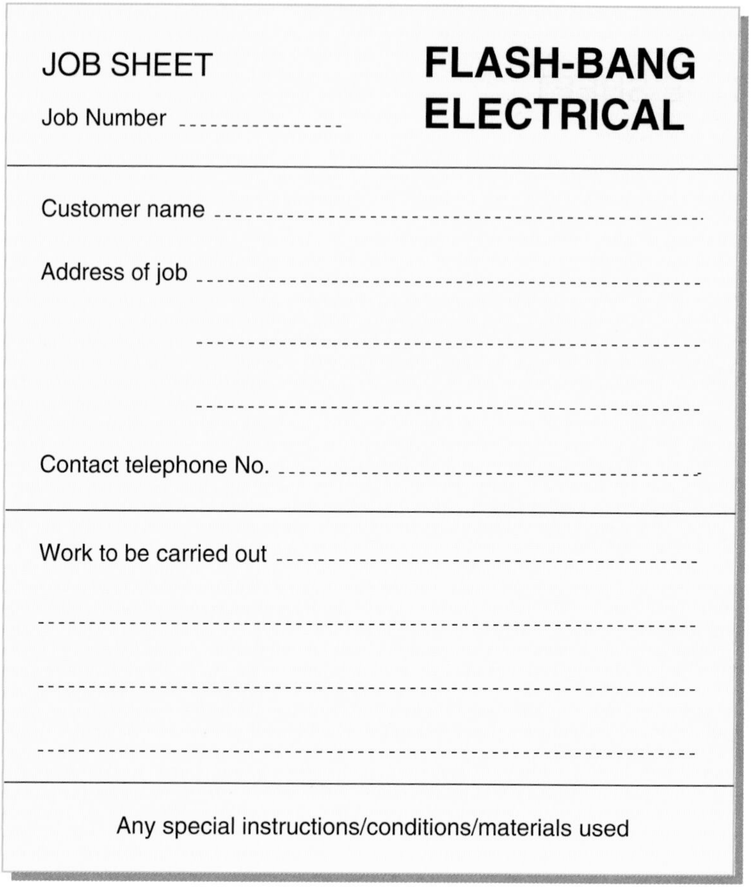

```
JOB SHEET                    FLASH-BANG
Job Number  ---------------  ELECTRICAL

Customer name  -------------------------------------------

Address of job  ------------------------------------------

                ------------------------------------------

                ------------------------------------------

Contact telephone No.  -----------------------------------


Work to be carried out  ----------------------------------

    -----------------------------------------------------

    -----------------------------------------------------

    -----------------------------------------------------

         Any special instructions/conditions/materials used
```

Figure 4.3 Job sheet.

confirm that goods have been delivered by the supplier, who will then send out an invoice requesting payment, usually at the end of the month.

The person receiving the goods must carefully check that all items stated on the delivery note have been delivered in good condition. Any missing or damaged items must be clearly indicated on the delivery note before signing because, by signing the delivery note, the person signing is saying 'yes, these items were delivered to me as my company's representative on that date and in good condition and I am now responsible for those goods'. Suppliers will replace materials damaged in transit, provided that they are notified within a set period, usually three days. The person receiving the goods should try to quickly determine their condition – has the packaging been damaged – does the container 'sound' like it might contain broken items? It is best to check at the time of delivery if possible or as soon as possible after delivery and within the notifiable period. Electrical

FLASH-BANG ELECTRICAL					DAYWORK SHEET

Client name _____

Job number/ref. _____

Date	Labour	Start time	Finish time	Total hours	Office use

Materials quantity	Description	Office use

Site supervisor or F.B. Electrical Representative responsible for carrying out work _____

Signature of person approving work and status e.g.

Client ☐   Architect ☐   Q.S. ☐   Main contractor ☐   Clerk of works ☐

Signature _____

Figure 4.4 Daywork sheet.

goods delivered to site should be handled carefully and stored securely until they are installed. Copies of delivery notes should be sent to Head Office so that payment can be made for the goods received.

Reports

On large jobs, the foreman or supervisor is often required to keep a report of the relevant events which happen on the site; for example, how many people from the company you work for are working on site each day; what goods were delivered; whether there were any breakages or accidents; and records of site meetings attended. Some firms have two separate documents, a site diary to record all daily events and a weekly report which is a summary of the week's events extracted from the site diary. The site diary remains on site and the weekly report is sent to Head Office to keep managers informed of the work's progress.

Electricity supply systems

The electricity supplies to houses, shops, offices and small industrial consumers is nominally set at 230 V single phase and 400 V three phase. The nominal voltage must be maintained by the supplier within a tolerance range of plus or minus ten per cent ($\pm$10%). So, a domestic supply must be maintained by the supplier between 207 V and 253 V for single phase supplies and between 360 V and 440 V for three phase supplies. All EU countries agreed to change to these values from 2005. The frequency is maintained at 50 cycles per second over 24 hours so that electric clocks remain accurate.

The electricity supply to domestic, commercial and small industrial consumers is usually protected at the incoming service cable position by a 100 A HBC fuse. Other items of equipment at this position are the energy meter to record the electricity consumption and a consumer unit/fuseboard to provide the protection for the final circuits and the earthing arrangements for the installation. An efficient and effective earthing system is essential to allow protective devices to operate quickly and effectively as described in Chapter 3 under the heading 'Protecting electrical equipment, circuits and people'.

Try this

To understand where electricity supplies come from, you should re-read the section 'Generation, transmission and distribution of electricity' in Chapter 3. You should also look at Figs 3.29 and 3.30 before starting on this section

The IEE Regulations (BS 7671) give details of the earthing arrangements in Chapter 54 and Figs 2.1 to 2.6. Five systems are described, but only three electricity supply systems are suitable for

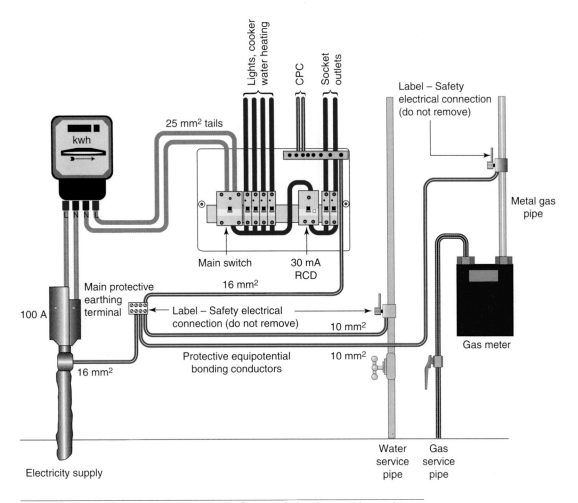

Figure 4.5 Cable sheath earth supplies (TN-S system) showing earthing and bonding arrangements.

public supplies and we will, therefore, only concern ourselves with these three supply systems.

Cable sheath earth supplies (TN-S system)

This is one of the most common types of supply system to be found in the UK, where the electricity company's supply is provided by underground cables. The neutral and protective conductors are separate throughout the system. The protective earth conductor is the metal sheath and armour of the underground cable and this is connected to the consumer's main earthing terminal. All extraneous conductive parts of the installation, gas pipes, water pipes and any lightning protective system are connected by protective equipotential bonding conductors to the main earthing terminal of the installation. The arrangement is shown in Fig. 4.5.

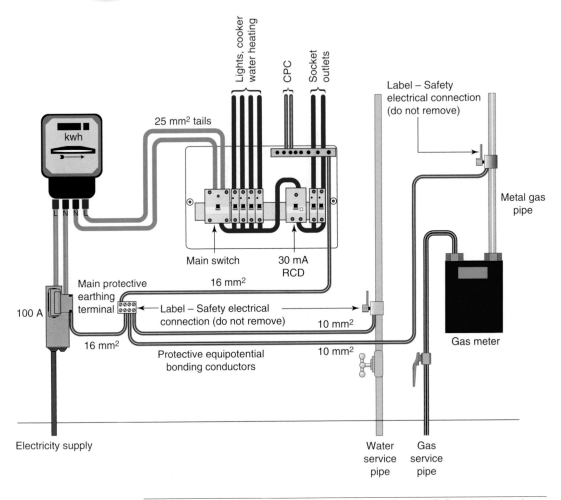

Figure 4.6 Protective multiple earthing (PME) supplies (TN-C-S system) showing earthing and bonding arrangements.

Protective multiple earthing supplies (TN-C-S system)

This type of underground supply is becoming increasingly popular to supply new installations in the UK. It is more commonly referred to as protective multiple earthing (PME). The supply cable uses a combined protective earth and neutral conductor. At the supply intake point a consumer's main earthing terminal is formed by the supply provider by connecting the earthing terminal to the neutral conductor. All extraneous conductive parts of the installation, gas pipes, water pipes and any lightning protective system are then connected to the main earthing terminals. Thus phase to earth faults are effectively converted into phase to neutral faults. The arrangement is shown in Fig. 4.6.

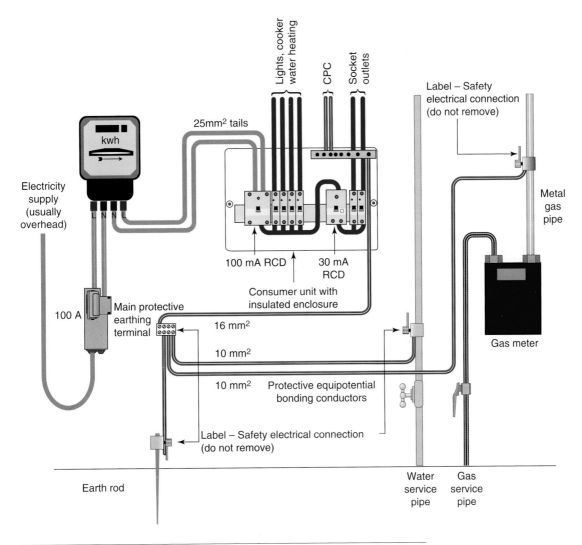

Figure 4.7 No earth provided supplies (TT system) showing earthing and bonding arrangements.

No earth provided supplies (TT system)

This is the type of supply more often found when the installation is fed from overhead cables. The supply authorities do not provide an earth terminal and the installation's circuit protective conductors must be connected to earth via an earth electrode provided by the consumer. An effective earth connection is sometimes difficult to obtain and in most cases a residual current device is provided when this type of supply is used. The arrangement is shown in Fig. 4.7.

Wiring and lighting circuits

The IEE's *On Site Guide* deals with the assumed current demand of points and states that for lighting outlets we should assume a current equivalent to a minimum of 100 W per lampholder. This means that for a domestic lighting circuit rated at 5 A or 6 A a maximum of 11 or 12 lighting outlets could be connected to each circuit. In practice, it is usual to divide the fixed lighting outlets into two or more circuits of seven or eight outlets each. In this way the whole installation is not plunged into darkness if one lighting circuit fails and it complies with IEE Regulation 314.1 which tells us to divide into circuits to minimize inconvenience and avoid danger.

Lighting circuits are usually wired in 1.0 mm or 1.5 mm cable using either a loop-in or joint-box method of installation. The loop-in method is universally employed with conduit installations or when access from above or below is prohibited after installation, as is the case with some industrial installations or blocks of flats. In this method the only joints are at the switches or lighting points, the live conductors being looped from switch to switch and the neutrals from one lighting point to another.

The use of junction boxes with fixed brass terminals is the method often adopted in domestic installations, since the joint boxes can be made accessible but are out of sight in the loft area or under floorboards. However, all connections must remain accessible for inspection, testing and maintenance (IEE Regulation 526.3).

The live conductors must be broken at the switch position in order to comply with the polarity regulations (IEE Regulation 612.7). A ceiling rose

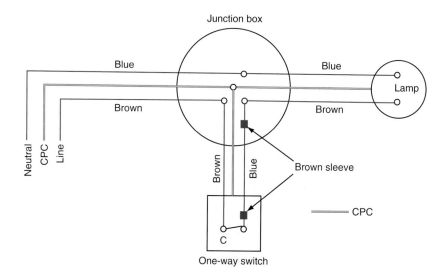

Figure 4.8 One-way switch control.

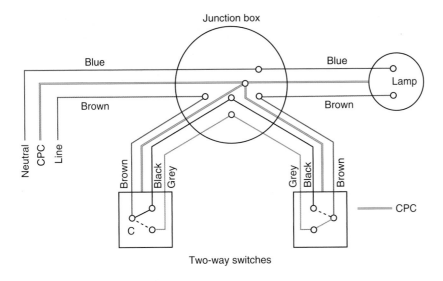

Figure 4.9 Two-way switch control.

may only be connected to installations operating at 250V maximum and must only accommodate one flexible cord unless it is specially designed to take more than one (IEE Regulations 559.6.1.2 and 3). Lampholders suspended from flexible cords must be capable of suspending the mass of the luminaire fixed to the lampholder (IEE Regulation 559.6.1.5).

The type of circuit used will depend upon the installation conditions and the customer's requirements. One light controlled by one switch is called one-way switch control (see Fig. 4.8).

A room with two access doors might benefit from a two-way switch control (see Fig. 4.9) so that the lights may be switched on or off at either position.

Fixing positions of switches and sockets

Part M of the Building Regulations requires switches and socket outlets in dwellings to be installed so that all persons, including those whose reach is limited, can easily reach them. The recommendation is that they should be installed in habitable rooms at a height of between 450mm and 1200mm from the finished floor level. This is shown in Fig. 4.10. The guidance given applies to all new dwellings but not to re-wires. However, these recommendations will undoubtedly 'influence' decisions taken when re-wiring dwellings.

Wiring and socket outlet circuits

Where a domestic appliance or mobile equipment is to be used, it should be connected by a plug to an adjacent or conveniently

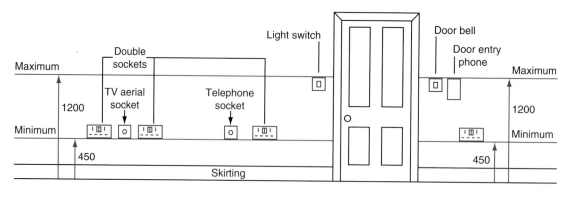

Figure 4.10 Fixing positions of switches and socket outlets.

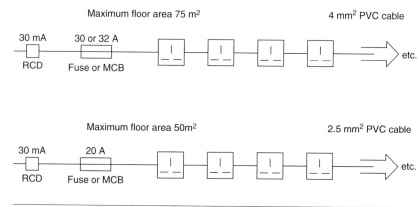

Figure 4.11 Block diagram of radial socket circuits.

accessible socket outlet, taking into account the length of the flexible cord normally fitted to portable appliances and mobile equipment (IEE Regulation 553.1.7). The length of flexible cord is usually no longer than two metres and so, pressing the plug into a socket outlet connects the appliance to the source of supply. Socket outlets therefore provide an easy and convenient method of connecting portable electrical appliances or mobile equipment to a source of supply.

Socket outlets can be obtained in 15, 13, 5 and 2 A ratings but the 13 A flat pin type complying with BS 1363 is the most popular for domestic installations in the United Kingdom. Each 13 A plug contains a cartridge fuse to give maximum potential protection to the flexible cord and the appliance which it serves.

Socket outlets may be wired on a ring or radial circuit and, in order that every appliance can be fed from an adjacent and convenient socket outlet, the number of sockets is unlimited provided that the floor area covered by the circuit does not exceed that given in Appendix 15 of the IEE Regulations and Figs 4.11 and 4.12 of this book.

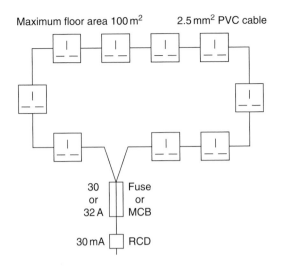

Maximum floor area 100 m²   2.5 mm² PVC cable

30 or 32 A — Fuse or MCB

30 mA — RCD

Figure 4.12 Block diagram of ring socket circuits.

Radial circuits

In a radial circuit each socket outlet is fed from the previous one. Live is connected to live, neutral and earth to earth at each socket outlet. The fuse and cable sizes are given in Appendix 15 of the IEE Regulations but circuits may also be expressed with a block diagram as shown in Fig. 4.11. The number of permitted socket outlets is unlimited but each radial circuit must not exceed the floor area stated and the known or estimated load.

Where two or more circuits are installed in the same premises, the socket outlets and permanently connected equipment should be reasonably shared out among the circuits so that the total load is balanced.

When designing ring or radial circuits, special consideration should be given to the loading in kitchens, which may require separate circuits. This is because the maximum demand of current-using equipment in kitchens may exceed the rating of the circuit cable and protection devices. Ring and radial circuits may be used for domestic or other premises where the demand of the current-using equipment is estimated not to exceed the rating of the protective devices for the chosen circuit.

Ring circuits

Ring circuits are very similar to radial circuits in that each socket outlet is fed from the previous one, but in ring circuits the last socket is wired back to the source of supply. Each ring final circuit conductor must be looped into every socket outlet or joint box which forms the ring and must be electrically continuous throughout its length. The number of permitted socket outlets is unlimited but each ring circuit must not cover more than 100 m² of floor area.

The circuit details are given in Appendix 15 of the IEE Regulations but may also be expressed by the block diagram given in Fig. 4.12.

Additional protection for socket outlets

Additional protection by 30 mA RCD is required in addition to overcurrent protection for all socket outlet circuits to be used by ordinary persons and intended for general use.

This additional protection is provided in case basic protection or fault protection fails or if the user of the installation is careless (IEE Regulation 415.1.1).

Try this

Definitions

Write in the margin here, or highlight in the text, a definition of:

- *a skilled person*
- *an instructed person*
- *an ordinary person*

An ordinary person is one who is neither an electrically skilled or instructed person.

Socket outlet numbers

The IEE Regulations allow us to install an unlimited number of socket outlets, the restriction being that each circuit must not exceed a given floor area as shown in Figs 4.11 and 4.12.

These days most households have lots of domestic appliances and electronic equipment, so how many sockets should be installed? Ultimately this is a matter for the customer and the electrical designer but most consumer organisations, the house builders NHBC and the Royal Society for the Prevention of Accidents (ROSPA) make the following general recommendations:

- The hard wiring for a single socket outlet is the same as the hard wiring for a double socket outlet. So, always install a double switched socket outlet unless there is a reason not to
- Kitchens will require between six and ten double sockets, fitted both above and below the work surface for specific appliances
- Utility room – two double sockets

Figure 4.13 Electrician installing socket outlet circuits.

- Sitting rooms will require between six and ten double sockets with one double socket situated next to any telephone outlet to power telecommunication equipment and two double sockets adjacent to the TV aerial outlet for TV, video and DVD supplies
- Double bedrooms – four to six double sockets
- Single bedrooms – four to six double sockets
- Hallways – two double sockets with one situated next to any telephone outlet
- Home office – six double sockets
- Garage – two double sockets

Try this

Socket numbers

- *how does your experience relate to the numbers recommended here?*
- *how many sockets are there in the rooms in your home?*
- *are there enough sockets at home?*
- *how many sockets do you install in houses where you work?*
- *why do you think there is a need for more socket outlets these days?*
- *make some not es in the margin here*

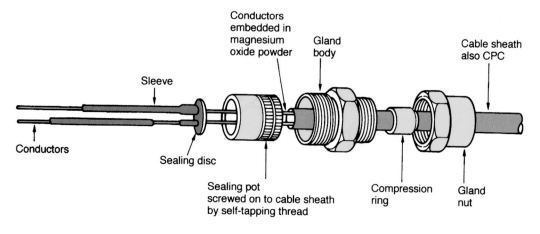

Figure 4.14 MI cable with terminating gland and seal.

Cables and enclosures

Power and lighting circuit conductors are contained within cables or enclosures. Part 5 of the IEE Regulations tells us that electrical equipment and materials must be chosen so that they are suitable for the installed conditions, taking into account temperature, the presence of water, corrosion, mechanical damage, vibration or exposure to solar radiation. Therefore, PVC insulated and sheathed cables are suitable for domestic installations but for a cable requiring mechanical protection and suitable for burying underground, a PVC/SWA cable would be preferable. These two types of cable are shown in Figs 2.2 and 2.3 in Chapter 2 of this book.

Mineral insulated (MI) cables are waterproof, heatproof and corrosion resistant with some mechanical protection. These qualities often make it the only cable choice for hazardous or high temperature installations such as oil refineries, chemical works, boiler houses and petrol pump installations. An MI cable with terminating gland and seal is shown in Fig. 4.14.

The FP 200 cable is another specialist cable. It is a fire-resistant cable, primarily intended for use in fire alarm and emergency lighting installations. Its appearance is very similar to an MI cable in that it is constructed as a thin pencil size tube but the outer sheath is made from a robust thermoplastic material and is much easier to terminate than an MI cable.

We will look at wiring enclosures in the next section but first let us look at the new wiring colours for all fixed wiring which came into force on 1 April 2006.

New wiring colours

On 31 March 2004 the IEE published Amendment No. 2 to BS 7671: 2001 which specified new cable core colours for all fixed wiring in United Kingdom electrical installations. These new core colours 'harmonize' the United Kingdom with the practice in mainland Europe.

Existing fixed cable core colours:

- **Single phase** - red line conductors, black neutral conductors and green combined with yellow for earth conductors
- **Three phase** - red, yellow and blue line conductors, black neutral conductors and green combined with yellow for earth conductors

    These core colours must not be used after 31 March 2006

New (harmonized) fixed cable core colours:

- **Single phase** - brown line conductors, blue neutral conductors and green combined with yellow for earth conductors
- **Three phase** - brown, black and grey line conductors, blue neutral conductors and green combined with yellow for earth conductors

    These core colours may be used from 31 March 2004

Extensions or alterations to existing *single phase* installations do not require marking at the interface between the old and new fixed wiring colours. However, a warning notice must be fixed at the consumer unit or distribution fuse board which states:

Caution – this installation has wiring colours to two versions of BS 7671. Great care should be taken before undertaking extensions, alterations or repair that all conductors are correctly identified.

Size of conductor

Appendix 4 of the IEE Regulations (BS: 7671) contains tables for determining the current carrying capacities of conductors. However, for standard domestic circuits, Table 4.1 gives a guide to cable size.

In this table, I am assuming a standard 230 V domestic installation, having a sheathed earth or PME supply terminated in a 100 A HBC fuse at the mains position. Final circuits are fed from a consumer unit, having Type B, MCB protection and wired in PVC insulated and sheathed cables with copper conductors having a grey thermoplastic PVC outer sheath or a white thermosetting cable with LSF (low smoke and fume properties). I am also assuming that the surrounding temperature throughout the length of the circuit does not exceed 30°C and the cables are run singly and clipped to a surface.

Table 4.1 Cable size for standard domestic circuits

Type of Final Circuit	Cable size (Twin and earth)	MCB rating, Type B (A)	Maximum floor area covered by circuit (m²)	Maximum length of cable run (m)
Fixed Lighting	1.0	6	–	40
Fixed Lighting	1.5	6	–	60
Immersion Heater	2.5	16	–	30
Storage Radiator	2.5	16	–	30
Cooker (oven only)	2.5	16	–	30
13 A Socket outlets (Radial circuit)	2.5	20	50	30
13 A Socket outlets (Ring circuit)	2.5	32	100	90
13 A Socket outlets (Radial circuit)	4.0	32	75	35
Cooker (oven and hob)	6.0	32	–	40
Shower (up to 7.5 kw)	6.0	32	–	40
Shower (up to 9.6 kw)	10	40	–	40

Wiring systems and enclosures

The final choice of a wiring system must rest with those designing the installation and those ordering the work, but whatever system is employed, good workmanship by competent persons and the use of proper materials is essential for compliance with the IEE Regulations (IEE Regulation 134.1.1). The necessary skills can be acquired by an electrical trainee who has the correct attitude and dedication to the craft.

PVC insulated and sheathed cable installations

PVC insulated and sheathed wiring systems are used extensively for lighting and socket installations in domestic dwellings. Mechanical damage to the cable caused by impact, abrasion, penetration, compression or tension must be minimized during installation (IEE Regulation 522.6.1). The cables are generally fixed using plastic clips incorporating a masonry nail, which means the cables can be fixed to wood, plaster or brick with almost equal ease. Cables should be run horizontally or vertically, not diagonally, down a wall. All kinks should be removed so that the cable is run straight and neatly between clips fixed at equal distances providing adequate support for the cable so that it does not become damaged by its own weight, as shown in Table 4.2

Table 4.2 Spacing of cable supports. Reproduced from the IEE *On Site Guide* by kind permission of the Institution of Electrical Engineers

Table 4A Spacings of supports for cables in accessible positions

Overall diameter of cable*	Maximum spacings of clips							
	Non-armoured thermosetting, thermoplastic or lead sheathed cables				Armoured cables		Mineral insulated copper sheathed or aluminium sheathed cables	
	Generally		In caravans					
	Horizontal†	Vertical†	Horizontal†	Vertical†	Horizontal†	Vertical†	Horizontal†	Vertical†
	2	3	4	5	6	7	8	9
mm	mm	mm	mm	mm	mm	mm	mm	mm
Not exceeding 9	250	400	250 (for all sizes)	400 (for all sizes)	–	–	600	800
Exceeding 9 and not exceeding 15	300	400			350	450	900	1200
Exceeding 15 and not exceeding 20	350	450			400	550	1500	2000
Exceeding 20 and not exceeding 40	400	550			450	600	–	–

Note: For the spacing of supports for cables having an overall diameter exceeding 40mm, and for single-core cables having conductors of cross-sectional area 300mm² and larger, the manufacturer's recommendations should be observed.

*For flat cables taken as the dimension of the major axis.

†The spacings stated for horizontal runs may be applied also to runs at an angle of more than 30 from the vertical. For runs at an angle of 30° or less from the vertical, the vertical spacings are applicable.

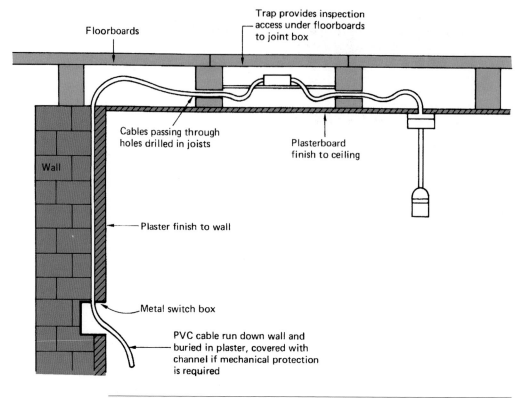

Figure 4.15 A concealed PVC sheathed wiring system.

(IEE Regulation 522.8.4). Where cables are bent, the radius of the bend should not cause the conductors to be damaged.

Terminations or joints in the cable may be made in ceiling roses, junction boxes or behind sockets or switches, provided that they are enclosed in a non-ignitable material, are properly insulated and are mechanically and electrically secure. All joints must be accessible for inspection and maintenance when the installation is completed (IEE Regulation 526.3).

Where PVC insulated and sheathed cables are concealed in walls, floors or partitions, they must be provided with a box incorporating an earth terminal at each outlet position. Figure 4.15 shows a typical concealed PVC sheathed wiring system.

To identify the most probable cable routes, IEE Regulation 522.6.6 tells us that outside a zone formed by a 150 mm border all around a wall edge, cables can only be run horizontally or vertically to a point or accessory unless they are contained in a substantial earthed enclosure such as a conduit, which can withstand nail penetration, as shown in Fig. 4.16. Where this protection cannot be complied

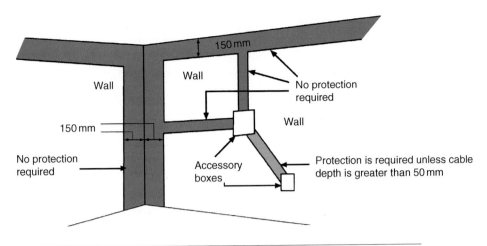

Figure 4.16 Permitted cable routes.

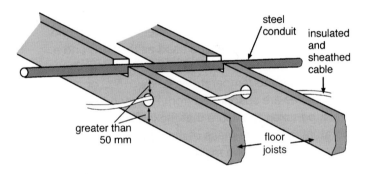

Notes:
1. Maximum diameter of hole should be 0.25 x joist depth.
2. Holes on centre line in a zone between 0.25 and 0.4 x span.
3. Maximum depth of notch should be 0.125 x joist depth.
4. Notches on top in a zone between 0.1 and 0.25 x span.
5. Holes in the same joist should be at least 3 diameters apart.

Figure 4.17 Correct installation of cables in floor joists.

with, then additional 30 mA RCD protection must be provided if the installation is to be used by ordinary people (IEE Regulation 522.6.7).

Where holes are drilled in floor joists to accommodate cable runs, they must meet the requirements shown in Fig. 4.17.

Conduit installations

A conduit is a tube, channel or pipe in which insulated conductors are contained. The conduit, in effect, replaces the PVC outer sheath of a cable, providing mechanical protection for the insulated conductors. A conduit installation can be re-wired easily or altered at anytime, and this flexibility, coupled with mechanical protection, makes conduit installations popular for commercial and industrial applications.

There are three types of conduit used in electrical installation work: steel, PVC and flexible.

Steel conduit

Steel conduit offers the conductors within a great deal of protection from mechanical damage. Steel conduit installations, therefore, find an application in industrial environments.

PVC conduit

PVC conduit used on typical electrical installations is heavy gauge standard impact tube manufactured to BS 4607. The conduit size and range of fittings are the same as those available for metal conduit. PVC conduit is most often joined by placing the end of the conduit into the appropriate fitting and fixing with a PVC solvent adhesive. PVC conduit can be bent by hand using a bending spring of the same diameter as the inside of the conduit.

The advantages of a PVC conduit system are that it can be installed much more quickly than steel conduit and is non-corrosive, but it does not have the mechanical strength of steel conduit. Since PVC conduit is an insulator it cannot be used as the CPC and a separate conductor must be run to every outlet. It is not suitable for installations subjected to temperatures below −5°C or above 60°C. Where luminaires are suspended from PVC conduit boxes, precautions must be taken to ensure that the lamp does not raise the box temperature or that the mass of the luminaire supported by each box does not exceed the maximum recommended by the manufacturer (IEE Regulations 522.1 and 2). PVC conduit also expands much more than metal conduit and so long runs require an expansion coupling to allow for conduit movement and help to prevent distortion during temperature changes.

All conduit installations must be erected first before any wiring is installed (IEE Regulation 522.8.2).

A limit must be placed on the number of bends between boxes in a conduit run and the number of cables which may be drawn into a conduit to prevent the cables being strained during wiring. Tables 5.2 and 5.3 of *Basic Electrical Installation Work* 6th Edition (ISBN 9780080966281) give a guide to the cable capacities of conduits and trunking.

Flexible conduit

Flexible conduit is made of interlinked metal spirals often covered with a PVC sleeving. The tubing must not be relied upon to provide a continuous earth path and, consequently, a separate CPC must be run either inside or outside the flexible tube (IEE Regulation 543.2.1).

Flexible conduit is used for the final connection to motors so that the vibrations of the motor are not transmitted throughout the electrical installation and to allow for modifications to be made to the final position of the motor for drive belt adjustments.

Trunking installations

A trunking is an enclosure provided for the protection of cables which is normally square or rectangular in cross-section, having one removable side. Trunking may be thought of as a more accessible conduit system and for industrial and commercial installations it is replacing the larger conduit size. A trunking system can have great flexibility when used in conjunction with conduit; the trunking forms the background or framework for the installation, with conduits running from the trunking to the point controlling the current using apparatus.

Trunking is supplied in 3 m lengths and various cross-sections measured in millimetres from 50 × 50 up to 300 × 150. Most trunking is available in either steel or plastic.

Metallic trunking

Metallic trunking is formed from mild steel sheet, coated with grey or silver enamel paint for internal use or a hot-dipped galvanized coating where damp conditions might be encountered. A wide range of accessories are available, such as 45° bends, 90° bends, tee and four-way junctions for speedy on-site assembly. Alternatively, bends may be fabricated in lengths of trunking, as shown in Fig. 4.18. This may be necessary or more convenient if a bend or set is non-standard, but it does take more time to fabricate bends than merely to bolt on standard accessories.

Cable tray installations

Cable tray is a sheet-steel channel with multiple holes. The most common finish is hot-dipped galvanized but PVC coated tray is also available. It is used extensively on large industrial and commercial installations for supporting MI and SWA cables which are laid on the cable tray and secured with cable ties through the tray holes.

Cable tray should be adequately supported during installation by brackets which are appropriate for the particular installation. The tray should be bolted to the brackets with round-headed bolts and nuts, with the round head inside the tray so that cables drawn along the tray are not damaged.

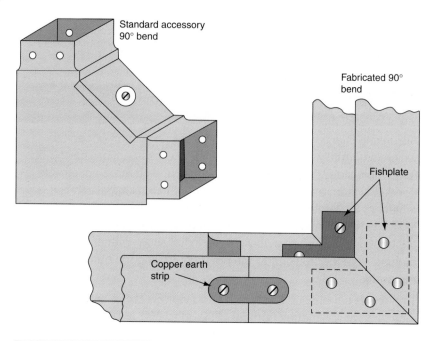

Figure 4.18 Alternative trunking bends.

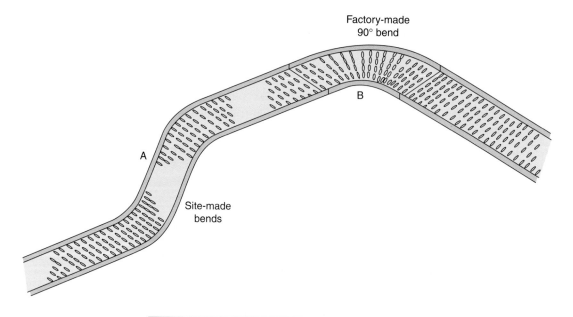

Figure 4.19 Cable tray with bends.

The tray is supplied in standard widths from 50 mm to 900 mm and a wide range of bends, tees and reducers is available. Figure 4.19 shows a factory-made 90° bend at B. The tray can also be bent using a cable tray bending machine to create bends such as that shown at A in Fig. 4.19.

Cable basket installations

Cable basket is becoming very popular for industrial installations. It is made from steel wire into a basket channel with sides.

Cable basket allows maximum airflow around the cables which are laid into the basket without fixing.

Cable basket requires similar installation techniques to cable tray and should be adequately supported.

PVC/SWA installations

A PVC/SWA cable is shown in Fig. 2.3.

MI cable installations

Mineral insulated cables are available with bare sheaths or with a PVC oversheath. The cable sheath provides sufficient mechanical protection for all but the most severe situations, where it may be necessary to fit a steel sheath or conduit over the cable to give extra protection, particularly near floor level in some industrial situations. (Figure 4.14 shows an MI cable termination.)

The cable may be laid directly in the ground, in ducts, on cable tray or clipped directly to a structure. It is not affected by water, oil or the cutting fluids used in engineering and can withstand very high temperatures or even fire. The cable diameter is small in relation to its current carrying capacity and it should last indefinitely if correctly installed because it is made from inorganic materials. These characteristics make the cable ideal for emergency circuits, boiler-houses, furnaces, petrol stations and chemical plant installations.

Special installations

All electrical installations and installed equipment must be safe to use and free from the dangers of electric shock, but some installations require special consideration because of the inherent dangers of the installed conditions. The danger may arise because of the corrosive or explosive nature of the atmosphere, because the installation must be used in damp or low temperature conditions or because there is a need to provide additional mechanical protection for the electrical system. In this section we will consider some of the installations which require special consideration.

Bathroom installations

In rooms containing a fixed bath tub or shower basin, additional regulations are specified. This is to reduce the risk of electric shock to people in circumstances where body resistance is lowered because of contact with water. The regulations can be found in Section 701 and can be summarized as follows:

- socket outlets MUST NOT be installed and no provision is made for the connection of portable appliances unless the socket can be fitted 3 metres horizontally beyond the zone 1 boundary within the bath or shower room (IEE Regulation 701.512.4).
- only shaver sockets which comply with BS EN 60742, that is those which contain an isolating transformer, may be installed in zone 2 or outside the zones in the bath or shower room (IEE Regulation 701.512.4).
- all circuits in the bath or shower room, both power and lighting, must be additionally protected by an RCD having a rated maximum operating time of 30 mA (IEE Regulation 701.411.3.3).
- there are restrictions to where appliances, switchgear and wiring accessories may be installed. See the next section on 'Zones for bath and shower rooms'.

Figure 4.20 Bathroom installations require special consideration.

- local supplementary equipotential bonding (IEE Regulation 701.415.2) must be provided to all gas, water and central heating pipes in addition to metallic baths, UNLESS THE FOLLOWING TWO REQUIREMENTS ARE BOTH IN PLACE:

  i all bathroom circuits, both power and lighting are additionally protected by 30 mA RCD and

  ii the bath or shower room is located in a building which has protective equipotential bonding in place as shown in Figs 4.5 to 4.7 earlier in this chapter (IEE Regulation 411.3.1.2).

Note: Local supplementary equipotential bonding may be an ADDITIONAL REQUIREMENT of the Local Authority regulations in, for example, licensed premises, student accommodation and rented property.

Zones for bath and shower rooms

Locations that contain a bath or shower are divided into zones or separate areas as shown in Fig. 4.21.

Zone 0 – the bath tub or shower basin itself, which can contain water and is, therefore, the most dangerous zone

Zone 1 – the next most dangerous zone in which people stand in water

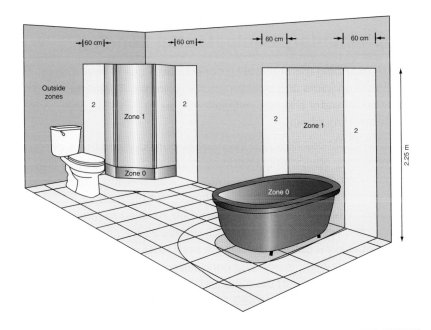

Figure 4.21 Cross-section through bathroom showing zones.

Zone 2 – the next most dangerous zone in which people might be in contact with water

Outside the Zones - people are least likely to be in contact with water but are still in a potentially dangerous environment

Electrical equipment and accessories are restricted within the zones:

Zone 0 – being the most potentially dangerous zone, for all practical purposes no electrical equipment can be installed in this zone. However, the IEE Regulations permit that, where SELV fixed equipment not exceeding 12 V cannot be located elsewhere, it may be installed in this zone

Zone 1 – water heaters, showers and shower pumps and SELV fixed equipment

Zone 2 – luminaires, fans and heating appliances and equipment from zone 1 plus shaver units to BS EN 60742

Outside the Zones - fixed appliances are allowed plus the equipment from zones 1 and 2, appliances are allowed plus accessories except socket outlets unless the room is very large and the socket outlet can be installed 3 m beyond the zone 1 boundary

If underfloor heating is installed in these areas it must have an overall earthed metallic grid or the heating cable must have an earthed metallic sheath which is connected to the protective conductor of the supply (IEE Regulation 701.753).

Supplementary equipotential bonding

Modern plumbing methods make considerable use of non-metals (PTFE tape on joints and plastic pipe for example). Therefore, the metalwork of water and gas installations cannot be relied upon to be continuous throughout.

The IEE Regulations describe the need to consider additional or supplementary equipotential bonding in situations where there is a high risk of electric shock (for example, in kitchens and bathrooms). In rooms containing a fixed bath or shower, supplementary equipotential bonding conductors **must** be installed to reduce to a minimum the risk of an electric shock **unless** all the circuits in the room containing a bath or shower are both RCD protected AND protective equipotential bonding conductors are connected to the main earthing terminal of the electrical installation as described at the beginning of this section (IEE Regulations 415.2, 411.3.2 and 701.415.2).

The supplementary equipotential bonding conductors in domestic premises will normally be 4.0 mm^2 copper, and connection must

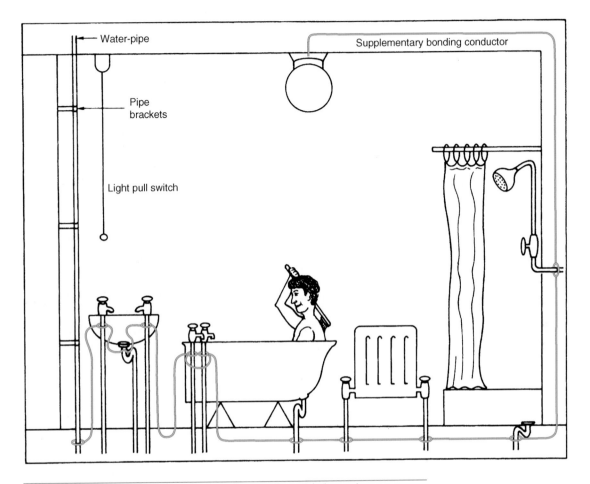

Figure 4.22 Supplementary bonding in bathrooms to metal pipework.

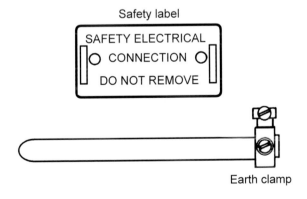

Figure 4.23 Typical earth bonding clamp.

be made to a cleaned pipe, using a suitable bonding clip. Fixed at or near the connection must be a permanent label saying '**Safety electrical connection - do not remove**' (IEE Regulation 514.3) as shown in Fig. 4.23.

Figure 4.24 Construction sites require special consideration.

Temporary installations (construction sites)

Temporary electrical supplies provided on construction sites can save many man-hours of labour by providing the energy required for fixed and portable tools and lighting which speeds up the completion of a project. However, construction sites are dangerous places and the temporary electrical supply which is installed to assist the construction process must comply with all of the relevant wiring regulations for permanent installations (IEE Regulation 110.1). All equipment must be of a robust construction in order to fulfil the on-site electrical requirements while being exposed to rough handling, vehicular nudging, the wind, rain and sun. All equipment, socket outlets, plugs and couplers must be of the industrial type to BS EN 60309 and specified by IEE Regulation 704.511.1 as shown in Fig. 4.25.

IEE Regulation 704 tells us that reduced low voltage is 'strongly prefered' for mobile tools, handlamps, plant and equipment on construction sites. Construction site supplies would typically be:

- 400 V three phase for supplies to major items of plant having a rating above 3.75 kW such as cranes and lifts. These supplies must be wired in armoured cables

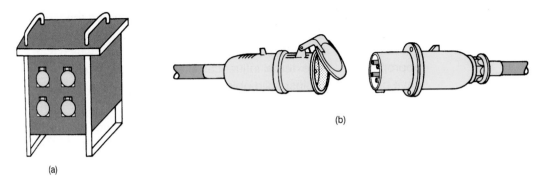

(b)

(a)

Figure 4.25 110 V distribution unit and cable connectors.

- 230 V single phase for supplies to items of equipment which are robustly installed such as flood-lighting towers, small hoists and site offices. These supplies must be wired in armoured cable unless run inside the site offices
- 110 V single phase for supplies to all mobile hand tools and lighting equipment. The supply is usually provided by a reduced voltage distribution unit which incorporates splash-proof sockets fed from a centre-tapped 110 V transformer. This arrangement limits the voltage to earth to 55 V, which is recognized as safe in most locations. A 110 V distribution unit is shown in Fig. 4.25. Edison screw lamps are used for 110 V lighting supplies so that they are not interchangeable with 230 V site office lamps.

There are occasions when even a 110 V supply from a centre-tapped transformer is too high; for example, supplies to inspection lamps for use inside damp or confined places. In these circumstances a safety extra-low voltage (SELV) supply would be required.

Industrial plugs have a keyway which prevents a tool from one voltage being connected to the socket outlet of a different voltage. They are also colour coded for easy identification as follows:

    440 V – red
    230 V – blue
    110 V – yellow
     50 V – white
     25 V – violet

Agricultural and horticultural installations

Especially adverse installation conditions are to be encountered on farms and in commercial greenhouses because of the presence of livestock, vermin, dampness, corrosive substances and mechanical damage. The 17th Edition of the IEE Wiring Regulations consider these installations very special locations and has devoted the whole of Section 705 to their requirements. In situations accessible to livestock the electrical equipment should be of a type which is appropriate for the external influences likely to occur and should have protection against solid objects and water splashing from any direction (IEE Regulation 705.512.2).

Horses and cattle have a very low body resistance, which makes them susceptible to an electric shock at voltages lower than 25 V rms.

In buildings intended for livestock, all fixed wiring systems must be inaccessible to the livestock and cables liable to be attacked by vermin must be suitably protected.

PVC cables enclosed in heavy duty PVC conduit are suitable for installations in most agricultural buildings. All exposed and extraneous metalwork must be provided with supplementary equipotential bonding in areas where livestock is kept (IEE Regulation 705.415.2.1). In most situations, waterproof socket outlets to BS 196 must be installed. All socket outlet circuits must be protected by an RCD complying with the appropriate British Standard and the operating current must not exceed 30 mA.

Figure 4.26 Farms and commercial greenhouses require special consideration.

Cables buried on agricultural or horticultural land should be buried at a depth of not less than 600 mm, or 1000 mm where the ground may be cultivated, and the cable must have an armour sheath and be further protected by cable tiles. Overhead cables must be insulated and installed so that they are clear of farm machinery or placed at a minimum height of 6.0 m to comply with IEE regulation 705.522.

Hazardous area installations

Most flammable liquids only form an explosive mixture between certain concentration limits. Above and below this level of concentration the mix will not explode. The lowest temperature at which sufficient vapour is given off from a flammable substance to form an explosive gas-air mixture is called the *flash-point*. A liquid which is safe at normal temperatures will require special consideration if heated to flash-point. An area in which an explosive gas-air mixture is present is called a *hazardous area*, as defined by BS EN 60079 and BS EN 50014: 1998, and any electrical apparatus or equipment within a hazardous area must be classified as flameproof. Flameproof equipment is manufactured to a robust standard of construction. All access and connection points have wide machined flanges which damp the flame in its passage across the flange. Flanged surfaces are firmly bolted together with many recessed bolts, as shown in Fig. 4.27. Wiring systems within a hazardous area must be flameproof fittings using an appropriate method such as:

- Mineral insulated cables terminated into accessories with approved flameproof glands. These have a longer gland thread than normal MICC glands of the type shown in Fig. 4.14. Where the cable is laid underground it must be protected by a PVC sheath and laid at a depth of not less than 500 mm

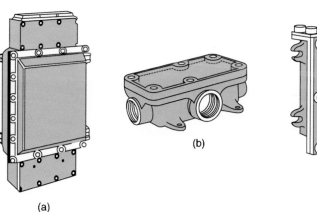
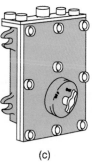
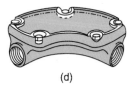

(a)

(b)

(c)

(d)

Figure 4.27 Flameproof fittings.

Figure 4.28 Flameproof equipment markings.

- PVC armoured cables terminated into accessories with approved flameproof glands or any other wiring system which is approved by the British Standard. All certified flameproof enclosures will be marked **Ex**, indicating that they are suitable for potentially explosive situations, or **EEx**, where equipment is certified to the harmonized European Standard. All the equipment used in a flameproof installation must carry the appropriate markings, as shown in Fig. 4.28 if the integrity of the wiring system is to be maintained. Flammable and explosive installations are to be found in the petroleum and chemical industries, which are classified as group II industries. Mining is classified as group I and receives special consideration from the Mining Regulations because of the extreme hazards of working underground. Petrol filling pumps must be wired and controlled by flameproof equipment to meet the requirements of the Petroleum Regulation Act 1928 and 1936 and any local licensing laws concerning the keeping and dispensing of petroleum spirit.

Try this

Definitions

Write down in the margin or highlight in the text a definition for:
- *flash point*
- *hazardous area*

Support and fixing methods for electrical equipment

Individual conductors may be installed in trunking or conduit and individual cables may be clipped directly to a surface or laid on a tray using the wiring system which is most appropriate for the particular installation. The installation method chosen will depend upon the contract specification, the fabric of the building and the type of installation - domestic, commercial or industrial.

Figure 4.29 Methods of fixing electrical equipment.

It is important that the wiring systems and fixing methods are appropriate for the particular type of installation and compatible with the structural materials used in the building construction. The electrical installation must be compatible with the installed conditions, must not damage the fabric of the building or weaken load-bearing girders or joists.

Let us look at some of the methods of fixing electrical cables and equipment.

Cable clips

PVC insulated and sheathed wiring systems are usually fixed with PVC clips in order to comply with IEE Regulations 522.8.3 and 4 and the table shown earlier in this chapter (Table 4.2). The clips are supplied in various sizes to hold the cable firmly and the fixing nail is a hardened masonry nail. Figure 4.30 shows a cable clip of this type.

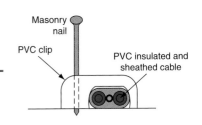

Figure 4.30 PVC insulated and sheathed cable clip.

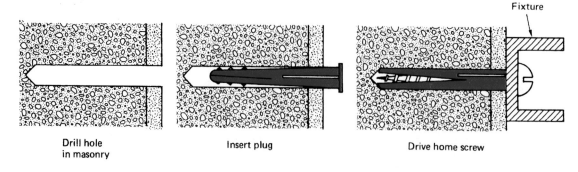

Drill hole
in masonry

Insert plug

Drive home screw

Figure 4.31 Screw fixing to plastic plug.

The use of a masonry nail means that fixings to wood, plaster, brick or stone can be made with equal ease.

When heavier cables, trunking, conduit or luminaires have to be fixed, a screw fixing is often needed. Wood screws may be screwed directly into wood but when fixing to brick, stone, plaster or concrete it is necessary to drill a hole in the masonry material, which is then plugged with a material to which the screw can be secured.

Plastic plugs

A plastic plug is made of a hollow plastic tube split up to half its length to allow for expansion. Each size of plastic plug is colour coded to match a wood screw size.

A hole is drilled into the masonry, using a masonry drill of the same diameter and to the same length as the plastic plug (see Fig. 4.31). The plastic plug is inserted into the hole and tapped home until it is level with the surface of the masonry. Finally the fixing screw is driven into the plastic plug until it becomes tight and the fixture is secure.

Expansion bolts

The most well known expansion bolt is made by Rawlbolt and consists of a split iron shell held together by a steel ferrule at one end and a spring wire clip at the other end. Tightening the bolt draws up an expanding bolt inside the split iron shell, forcing the iron to expand and grip the masonry. Rawlbolts are for heavy duty masonry fixings (see Fig. 4.32).

A hole is drilled in the masonry to take the iron shell and ferrule. The iron shell is inserted with the spring wire clip end first so that the

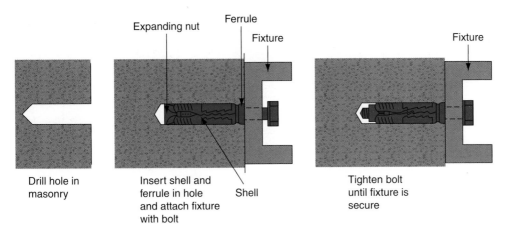

Drill hole in masonry

Insert shell and ferrule in hole and attach fixture with bolt — Shell

Tighten bolt until fixture is secure

Figure 4.32 Expansion bolt fixing.

ferrule is at the outer surface. The bolt is passed through the fixture, located in the expanding nut and tightened until the fixing becomes secure.

For the most robust fixing to masonry material an expansion bolt, such as that made by Rawlbolt, should be used.

Spring toggle bolts

A spring toggle bolt provides one method of fixing to hollow partition walls which are usually faced with plasterboard and a plaster skimming. Plasterboard and plaster wall or ceiling surfaces are not strong enough to support a load fixed directly into the plasterboard, but the spring toggle spreads the load over a larger area, making the fixing suitable for light loads (see Fig. 4.33).

A hole is drilled through the plasterboard and into the cavity. The toggle wings are compressed and passed through the hole in the plasterboard and into the cavity where they spring apart and rest on the cavity side of the plasterboard. The bolt is tightened until the fixing becomes firm.

Girder fixings

In many commercial and industrial buildings it is necessary to fix trunking, conduit and tray to the structural fabric of the building. In general, it is unacceptable to drill holes in the load-bearing structure of the building to support the electrical installation for fear

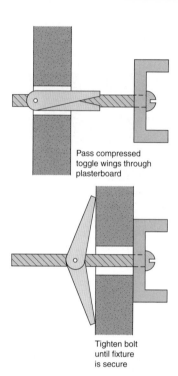

Pass compressed
toggle wings through
plasterboard

Tighten bolt
until fixture
is secure

Figure 4.33 Spring toggle bolt fixing.

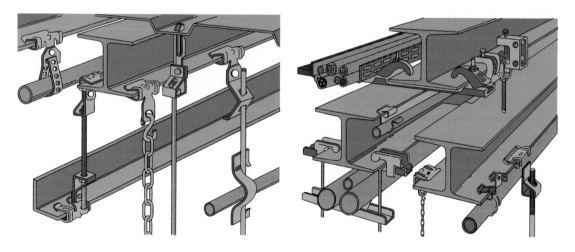

Figure 4.34 Girder supports.

of weakening the building structure itself. However, spring clips
or compression brackets are available which attach to the girders
and hold the electrical systems securely. Figure 4.34 shows some
manufactured girder supports for electrical equipment.

Electrical installation, inspection and testing

Having fixed everything securely and completed the electrical installation, it must be inspected and tested before being put into operation. The process of inspection is a visual thing. The installation must be carefully scrutinized before being tested to ensure that it is safe to be made electrically 'alive'. The process of testing implies the use of instruments to obtain readings. The test results must be compared with 'relevant criteria' to make sure that they are satisfactory (IEE Regulations 610 to 634).

The tests required by the IEE Regulations BS:7671 Requirements for Electrical Installations, must be carried out in the order given below so that safety systems are tested first. If any test indicates a failure to comply, then that test and all preceding tests must be repeated after the fault has been rectified.

1. Continuity of protective conductors including main and supplementary equipotential bonding

The objective of the test is to ensure that every circuit protective conductor is correctly connected and has a very low resistance.

The test is made with the supply disconnected, from the consumer's earthing terminal to the farthest point of each CPC and each protective equipotential bonding conductor as shown in Fig. 4.35 using an ohmmeter continuity tester. The resistance of the long test lead is subtracted from the test readings to give the resistance value of the CPC.

Relevant criteria tell us that a satisfactory test result would be resistance values in the order of $0.05\,\Omega$ or less (IEE Guidance Note 3).

2. Continuity of ring final circuit conductors

This test is carried out with the supply disconnected using an ohmmeter and verifies the continuity of the phase neutral and protective conductors. It also verifies that the conductors are all connected in a 'ring' and that the ring has no breaks or interconnections.

3. Insulation resistance

The object of the test is to verify the 'quality' of the insulation and that the insulation resistance has a very high value. The test is made at the consumer unit with the supply disconnected using an insulation resistance meter which supplies a voltage of 500 V.

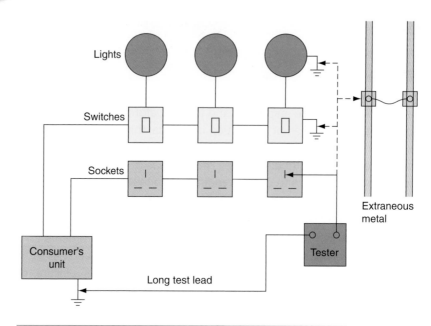

Lights

Switches

Sockets

Consumer's unit

Long test lead

Extraneous metal

Tester

Figure 4.35 Testing continuity of protection conductors.

Pilot indicator lamps, discharge lighting and electronic equipment must be temporarily disconnected before this test begins to avoid false readings and possible damage to equipment as a result of the test voltage.

Relevant criteria tell us that a satisfactory test result would be a minimum resistance value of $1.0\,M\Omega$ but if values of less than $2\,M\Omega$ are recorded then this might indicate a latent but not yet visible fault in the installation which would require further investigation. A new installation would typically have an insulation resistance value of infinity (symbol ∞).

4. Polarity

The object of the test is to ensure that all fuses, MCBs and switches are connected in the line conductor only and that all socket outlets are correctly wired.

The test is carried out with the supply disconnected using an ohmmeter as follows:

1 switch off the supply at the main switch
2 remove all lamps and unplug all equipment
3 **fix a temporary link** between line and earth on the consumer's side of the main switch as shown in Fig. 4.36
4 test between the common terminal and earth at each switch connection
5 test between the live pin and earth at each socket outlet
6 **remove the link** when the test is completed.

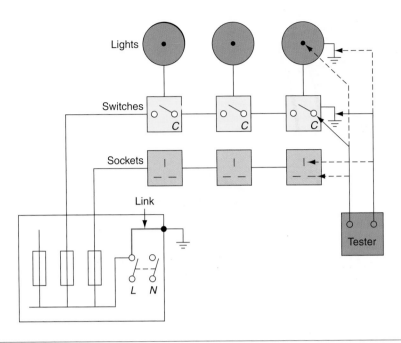

Figure 4.36 Testing polarity.

Relevant criteria tell us that a satisfactory test result would be a very low resistance value, approaching zero ohms for each test.

When all the tests are completed and proved satisfactory, the supply may be switched on. Functional testing is then carried out to ensure the correct operation of all circuits. Functional testing means that devices are operated to confirm that they are working properly and are correctly adjusted. The integral test button marked T or Test on an RCD should be pressed to prove the mechanical parts of the residual current device.

Electrical test instruments

Electrical installation testing in accordance with the relevant regulations demands that we use specialist test instruments. It is unacceptable for a professional electrician to carry out electrical testing using instruments bought at the local DIY superstore. Test instruments must meet the instrument standard BS EN 61557 and carry an 'in date' calibration certificate, otherwise test results are invalid.

Correct disposal of waste material

The Controlled Waste Regulations 1998 tell us that we have 'a duty of care' to handle, recover and dispose of waste responsibly.

The Environmental Protection (Duty of Care) Regulations 1991 tell us that any business has a duty to ensure that any waste produced is handled safely and in accordance with the law.

Figure 4.37 Disposing of waste material responsibly.

Your company is responsible for the waste that it produces even after handling it over to another party such as a skip hire company. If such a third party mishandles your waste or disposes of it irresponsibly, then it is the responsibility of the company you work for, not the skip hire company. The duty of care under the regulations has no 'time limit' and extends until the waste has either been finally and properly disposed of or fully recovered.

If a material has hazardous properties, it may need to be dealt with as 'Special Waste'. Containers may be classified as 'Special Waste' if they contain residues of hazardous or dangerous substances. If the residue is 'Special' then the whole container is 'Special Waste'.

Fluorescent lamps and tubes are now classified as hazardous waste. The environmentally responsible way to dispose of used lamps and tubes is to recycle them, probably through your electrical wholesaler.

Do not burn scrap cable on site, recycle it through a scrap metal merchant.

Electrotechnical companies produce very little waste material and even smaller amounts of 'Special Waste'. Most electrical contractors deal with waste by buying in the expertise and building in these costs to the total cost of a contract. However, this method still requires individuals to sort any waste responsibly by placing it in the appropriate skip or container.

To comply with the Waste Regulations:

- Make sure waste is transferred only to 'authorized' companies
- Make sure that the waste being taken is accompanied by the proper paperwork called 'waste transfer notes'
- Label waste skips and waste containers so that it is clear to everyone what type of waste is going into which skip or container
- Minimize the waste that you produce and do not leave it behind when a job is completed for someone else to clear away. As the producer of any waste, you are responsible for it. Remember there is no time limit on the duty of care for waste materials.

Key facts

If you have read and understood the whole of this chapter, you have completed most of the installing electrical equipment and systems outcomes of the City & Guilds 2357 knowledge units.

When you have completed the practical assessments required by the City & Guilds syllabus, which you are probably doing at your local college, you may be ready to tackle the on-line assessment. So, to prepare you for the on-line assessment, try the following assessment questions.

Assessment questions

Identify the statements as true or false. If only part of the statement is false, tick false

1 The Electricity at Work Regulations tell us that the IEE (called the IET from 31 March 2006) Wiring Regulations (BS 7671) is a code of practice which is widely recognised and accepted in the UK. If your electrical work meets the requirements of the IET Wiring Regulations, it will meet the requirements of all other relevant regulations.

True ☐ False ☐

2 The main reason for incorporating the Wiring Regulations into British Standards BS 7671 was to create harmonization with European Standards.

True ☐ False ☐

3 A cable sheath earth supply or TN-S System of supply is one of the most common types of underground supply in the UK.

True ☐ False ☐

4 A **delivery note** is a standard form completed by most electrical trainees to inform an employer of how much time has been spent working on a particular job.

True ☐ False ☐

5 When materials are delivered to site, the person receiving the goods is required to sign the driver's **time sheet** to prove that the supplier has delivered the goods as requested.

True ☐ False ☐

6 A radial socket circuit is wired from the source of supply to each socket in turn and the last socket is wired back to the source of supply.

True ☐ False ☐

7 From 1 April 2006 only the new wiring colours must be used for all fixed wiring. That is: brown for phase, blue for neutral and green combined with yellow for all single phase circuits.

True ☐ False ☐

8   Industrial installations use robust cable enclosures such as
conduits and trunking. A conduit is a square or rectangular
section from mild steel plate. A trunking is a tube, or pipe, in
which insulated conductors are contained.

True   ☐   False   ☐

9   Individual cables or accessories may be fixed directly to a
surface with a suitable nail, screw or bolt. A spring toggle bolt
provides a good method of fixing to concrete or masonry, a
Rawlbolt provides a good method of fixing to hollow partition
walls.

True   ☐   False   ☐

10   The 'Waste Regulations' tell us that we have a 'duty of care'
to handle, recover and dispose of waste responsibly. Your
company is responsible for the waste that it produces, so always
make sure that waste material is put into the proper skip and
taken away only by 'authorized' companies.

True   ☐   False   ☐

Multiple choice assessment questions

Tick the correct answer. Note that more than ONE answer may be correct

11 **The Electricity at Work Regulations are:**
- a. Non-statutory regulations ☐
- b. Statutory regulations ☐
- c. a code of practice ☐
- d. a British Standard ☐

12 **The IEE Regulations are:**
- a. Non-statutory regulations ☐
- b. Statutory regulations ☐
- c. a code of practice ☐
- d. a British Standard ☐

13 **A British Standard having a BS number is a:**
- a. Statutory regulation ☐
- b. Non-statutory regulation ☐
- c. British compliant standard ☐
- d. European harmonized standard ☐

14 **A British Standard having a BS EN number is a:**
- a. Statutory regulation ☐
- b. Non-statutory regulation ☐
- c. British compliant standard ☐
- d. European harmonized standard ☐

15 **Part 5 of the IEE Regulations deals with:**
- a. Protection for Safety ☐
- b. Selection and Erection of Equipment ☐
- c. Special Installations ☐
- d. Inspection and Testing ☐

16 **Part 6 of the IEE Regulations deals with:**

a. Protection for Safety ☐

b. Selection and Erection of Equipment ☐

c. Special Installations ☐

d. Inspection and Testing ☐

17 **A scale drawing showing the position of equipment by graphical symbols is a description of a:**

a. block diagram ☐

b. wiring diagram ☐

c. circuit diagram ☐

d. layout diagram or site plan ☐

18 **A diagram which shows the detailed connections between individual items of equipment is a description of a:**

a. block diagram ☐

b. wiring diagram ☐

c. circuit diagram ☐

d. layout diagram or site plan ☐

19 **A diagram which shows very clearly how a circuit works, where all components are represented by a graphical symbol is a description of a:**

a. block diagram ☐

b. wiring diagram ☐

c. circuit diagram ☐

d. layout diagram or site plan ☐

20 **A Time Sheet shows:**

a. a record of goods delivered by a supplier ☐

b. a record of work done which is outside the original contract ☐

c. information about work to be done, usually a small job ☐

d. the actual time spent working on a particular job or site ☐

21 **A Job Sheet or Job Card shows:**

a. a record of goods delivered by a supplier ☐

b. a record of work done which is outside the original contract ☐

c. information about work to be done, usually a small job ☐

d. the actual time spent working on a particular job or site ☐

22 **A Day Work Sheet shows:**

a. a record of goods delivered by a supplier ☐

b. record of work done which is outside the original contract ☐

c.   information about work to be done, usually a
     small job   ☐

d.   the actual time spent working on a particular job
     or site   ☐

23   A Delivery Note shows:

a.   a record of goods delivered by a supplier   ☐

b.   a record of work done which is outside the original
     contract   ☐

c.   information about work to be done, usually a
     small job   ☐

d.   the actual time spent working on a particular job
     or site   ☐

24   A cable sheath earth supply is also called a:

a.   TN-S system   ☐

b.   TN-C-S system   ☐

c.   TT system   ☐

d.   Standby system   ☐

25   A PME supply is also called a:

a.   TN-S system   ☐

b.   TN-C-S system   ☐

c.   TT system   ☐

d.   Standby system   ☐

26   A no earth provided supply is also called a:

a.   TN-S system   ☐

b.   TN-C-S system   ☐

c.   TT system   ☐

d.   Standby system   ☐

**27   The electricity supply to a domestic consumer is usually
      protected at the incoming service position by a:**

a.   Meter   ☐

b.   Double pole switch   ☐

c.   100 A MCB   ☐

d.   100 A HBC fuse   ☐

**28   The assumed current demand for each lighting point in
      a domestic installation should be based upon the
      equivalent of:**

a.   5 amps per lampholder   ☐

b.   6 amps per lampholder   ☐

c.   100 Watt per lampholder   ☐

d.   3 kW per lampholder   ☐

29 The protective Type B MCB for a lighting circuit fed from a consumer unit in 1.0 mm or 1.5 mm cable should be rated at:

a.   6 A or 10 A ☐

b.   10 A or 16 A ☐

c.   16 A or 32 A ☐

d.   32 A or 40 A ☐

30 The protective Type B MCB for a ring circuit fed from a consumer unit in 2.5 mm cable should be rated at:

a.   6 A or 10 A ☐

b.   10 A or 16 A ☐

c.   16 A or 32 A ☐

d.   32 A only ☐

31 Each ring circuit of 13 A sockets must cover a floor area of no more than:

a.   50 m² ☐

b.   75 m² ☐

c.   100 m² ☐

d.   unlimited ☐

32 A radial circuit of 13 A sockets wired in 2.5 mm PVC cable must cover a floor area of no more than:

a.   50 m² ☐

b.   75 m² ☐

c.   100 m² ☐

d.   unlimited ☐

33 A radial circuit of 13 A sockets wired in 4.0 mm PVC cable must cover a floor area of no more than:

a.   50 m² ☐

b.   75 m² ☐

c.   100 m² ☐

d.   unlimited ☐

34 An MI cable is especially suited to:

a.   domestic installations ☐

b.   fire alarm installations ☐

c.   burying underground ☐

d.   industrial installations ☐

35 A PVC/SWA cable is especially suited to:

a.   domestic installations ☐

b.   fire alarm installations ☐

    c.   burying underground ☐

    d.   industrial installations ☐

36 A PVC insulated and sheathed cable is especially suited to:

    a.   domestic installations ☐

    b.   fire alarm installations ☐

    c.   burying underground ☐

    d.   industrial installations ☐

37 FP 200 cables are especially suited to:

    a.   domestic installations ☐

    b.   fire alarm installations ☐

    c.   burying underground ☐

    d.   industrial installations ☐

38 A steel conduit installation is especially suited to:

    a.   domestic installations ☐

    b.   fire alarm installations ☐

    c.   burying underground ☐

    d.   industrial installations ☐

39 A metallic trunking installation is especially suited to:

    a.   domestic installations ☐

    b.   fire alarm installations ☐

    c.   burying underground ☐

    d.   industrial installations ☐

40 Cable tray installations are especially suited to:

    a.   domestic installations ☐

    b.   fire alarm installations ☐

    c.   burying underground ☐

    d.   industrial installations ☐

41 Bathroom installations receive special consideration in the IEE Regulations because of the hazard associated with:

    a.   electricity and flammable liquids ☐

    b.   electricity and water ☐

    c.   the presence of corrosive substances ☐

    d.   the potential for mechanical damage ☐

42 Construction sites receive special consideration in the IEE Regulations because of the hazard associated with:

    a.   electricity and flammable liquids ☐

    b.   electricity and water ☐

    c.   exposure to wind and rain ☐

    d.   presence of livestock and vermin ☐

43 **Agricultural installations receive special consideration in the IEE Regulations because of the hazard associated with:**

   a.   electricity and water   ☐

   b.   presence of livestock and vermin   ☐

   c.   potential for mechanical damage   ☐

   d.   electricity and flammable liquids   ☐

44 **Petrol pump installations receive special consideration from many statutory regulations because of the hazard associated with:**

   a.   electricity and water   ☐

   b.   electricity and flammable liquids   ☐

   c.   exposure to wind and rain   ☐

   d.   the temporary nature of the supply   ☐

45 **Locations containing a bath or shower are divided into zones or separate areas. The most dangerous zone is classified as:**

   a.   Zone 0   ☐

   b.   Zone 1   ☐

   c.   Zone 2   ☐

   d.   Zone 10   ☐

46 **The permissible colours of 230 V single phase fixed wiring up to 30 March 2006 was:**

   a.   brown, blue, green and yellow   ☐

   b.   brown, black, grey   ☐

   c.   red, black, green and yellow   ☐

   d.   red, yellow, green and yellow   ☐

47 **The new European harmonized fixed wiring colours which must be used after 1 April 2006 for a 230 V single phase circuit are:**

   a.   brown, blue, green and yellow   ☐

   b.   brown, black, grey   ☐

   c.   red, black, green and yellow   ☐

   d.   red, yellow, green and yellow   ☐

48 **PVC insulated and sheathed cables are very likely to be fixed and supported by:**

   a.   wood screws and plastic plugs   ☐

   b.   a PVC clip and hardened nail   ☐

   c.   an expansion bolt   ☐

   d.   a clip on girder fixing   ☐

49 **A lightweight piece of electrical equipment is very likely to be fixed to a plasterboard by:**

a. wood screws and plastic plugs ☐

b. a PVC clip and hardened nail ☐

c. an expansion bolt ☐

d. a spring toggle bolt ☐

50 **A heavy electric motor is very likely to be fixed to a concrete floor by:**

a. wood screws and plastic plugs ☐

b. a clip on girder fixing ☐

c. an expansion bolt ☐

d. a spring toggle bolt ☐

51 **A run of trunking suspended in an industrial installation is very likely to be fixed to the main structure of the building by:**

a. wood screws and plastic plugs ☐

b. a clip on girder fixing ☐

c. an expansion bolt ☐

d. a spring toggle bolt ☐

52 **A run of cable tray suspended in a modern supermarket building is very likely to be attached to the main structure of the building by:**

a. wood screws and plastic plugs ☐

b. a clip on girder fixing ☐

c. an expansion bolt ☐

d. a spring toggle bolt ☐

53 **What action is necessary to produce a 'secure electrical isolation'?**

a. Isolate the supply and observe that the voltage indicator reads zero ☐

b. First connect a test device such as a voltage indicator to the supply ☐

c. Larger pieces of equipment may require isolating at a local isolator switch ☐

d. The isolated supply must be locked off or secured with a small padlock ☐

54 A voltage proving unit:
    a.   is used for transmitting data along optical fibre cables ☐
    b.   provides a secure computer supply ☐
    c.   shows a voltage indicator to be working correctly ☐
    d.   tests for the presence of a mains voltage supply ☐

55 For working even a short distance above ground level for long periods, the safest piece of access equipment would be:
    a.   a stepladder ☐
    b.   a platform tower ☐
    c.   an extension ladder ☐
    d.   a hard hat ☐

56 An example of `Special Waste' is:
    a.   sheets of asbestos ☐
    b.   old fibre-glass roof insulation ☐
    c.   old fluorescent tubes ☐
    d.   part coils of PVC insulated cables ☐

57 Special Waste must be disposed of:
    a.   in the general site skips ☐
    b.   in the general site skip by someone designated to have a 'duty of care' ☐
    c.   at the 'Household Waste' recycling centre ☐
    d.   by an 'authorized company' using a system of waste transfer notes' ☐

58 The Health & Safety at Work Act places the responsibility for safety at work on:
    a.   the employer ☐
    b.   the employee ☐
    c.   both the employer and employee ☐
    d.   the main contractor ☐

59 Under the Health & Safety at Work Act an employer must ensure that:
    a.   the working conditions are appropriate and safety equipment is provided ☐
    b.   employees take reasonable care of themselves and others as a result of work activities ☐
    c.   employees co-operate with an employer and do not interfere with or misuse safety equipment ☐
    d.   that plant and equipment is properly maintained ☐

60 **Under the Health & Safety at Work Act employees must ensure that:**

   a.   the working conditions are appropriate and safety equipment is provided ☐

   b.   they take reasonable care of themselves and others as a result of work activities ☐

   c.   they co-operate with an employer and do not interfere with or misuse safety equipment ☐

   d.   plant and equipment is properly maintained ☐

CHAPTER 5

Environmental technology systems

This chapter describes the relevant environmental legislation and new environmental technology systems.

For many years now the United Kingdom has generated electricity in large power stations connected to the national grid and fuelled by coal, oil and nuclear energy.

We have recently become aware that the climate on planet earth has begun to change and respectable scientific sources tell us that this is being caused by CO_2 greenhouse gases that are emitted into the atmosphere.

The old coal powered power stations release a lot of CO_2, our oil reserves are very low and our nuclear power stations are reaching the end of their intended lifespan.

For this reason, and the fact that the United Kingdom has entered into a European-wide agreement to reduce CO_2 emissions, the government has been forced to consider how we will secure our electricity supply in the future.

New nuclear power stations will be built because they do not produce CO_2 but there is some fear about the safety of nuclear power.

New coal powered power stations will be built but these will incorporate CO_2 capture systems.

In the future we must reduce our dependence on oil because, as previously mentioned, we have few oil reserves of our own and friendly countries supplying us with oil now, such as Russia and Libya, may not be so friendly in the future.

Introduction to Electrical Installation Work. 978-0-08-096940-4

So, green energy, which produces no CO_2, looks very promising and the government has committed us to producing most of our future energy requirements for the national grid from offshore wind farms.

The government is also encouraging us as individuals to look at ways of reducing our CO_2 emissions by better insulation in our homes and producing green electrical energy for ourselves and the community in which we live. Other countries, such as Germany and Canada, are much further ahead with this than we are in the United Kingdom.

It is these new green technologies that we will look at in this chapter but let us first of all look at the laws and regulations that protect our environment.

Environmental laws and regulations

The environment describes the world in which we live, work and play; it relates to our neighbourhood and surroundings.

Environmental laws protect the environment in which we live by setting standards for the control of pollution to land, air and water.

If a wrong is identified in the area in which we now think of as 'environmental' it can be of two kinds:

1 An offence in common law which means damage to property, nuisance or negligence leading to a claim for damages.
2 A statutory offence against one of the laws dealing with the protection of the environment. These offences are nearly always 'crimes' and punished by fines or imprisonment rather than by compensating any individual.

The legislation dealing with the environment has evolved for each part – air, water, land, noise, radioactive substances. Where an organization's activities impact upon the environmental laws they are increasingly adopting environmental management systems which comply with ISO 14001. Let us now look at some of the regulations and try to see the present picture at the beginning of the new millennium.

Environmental Protection Act 1990

In the context of environmental law, the Environmental Protection Act 1990 was a major piece of legislation. The main sections of the act are:

Part 1　Integrated pollution control by HM Inspectorate of Pollution, and air pollution control by Local Authorities

Part 2　Wastes on land

Part 3   Statutory nuisances and clean air

Part 4   Litter

Part 5   Radioactive Substances Act 1960

Part 6   Genetically modified organisms

Part 7   Nature conservation

Part 8   Miscellaneous, including contaminated land.

The Royal Commission of 1976 identified that a reduction of pollutant to one medium (air, water or land) then led to an increase of pollutant to another. It, therefore, stressed the need to take an integrated approach to pollution control. The processes subject to an integrated pollution control are:

- Air emissions.
- Processes which give rise to significant quantities of special waste; that is, waste defined in law in terms of its toxicity or flammability.
- Processes giving rise to emissions to sewers or 'Red List' substances. These are 23 substances including mercury, cadmium and many pesticides, which are subject to discharge consent to the satisfaction of the Environment Agency.

Where a process is under integrated control, the Inspectorate is empowered to set conditions to ensure that the best practicable environmental option (BPEO) is employed to control pollution. This is the cornerstone of the Environmental Protection Act.

Pollution Prevention and Control Regulations 2000

The system of pollution prevention and control is replacing that of integrated pollution control established by the Environmental Protection Act 1990, thus bringing environmental law into the new millennium and implementing the European Directive (EC/96/61) on integrated pollution prevention and control. The new system was fully implemented in 2007.

Pollution prevention and control is a regime for controlling pollution from certain industrial activities. This regime introduces the concept of Best Available Technique (BAT) for reducing and preventing pollution to an acceptable level.

Industrial activities are graded according to their potential to pollute the environment:

- A(1) installations are regulated by the Environment Agency.
- A(2) installations are regulated by the Local Authority.
- Part B installations are also regulated by the Local Authority.

All three systems require the operators of certain industrial installations to obtain a permit to operate. Once an operator has submitted a permit application, the regulator then decides whether to issue a permit. If one is issued it will include conditions aimed at reducing and preventing pollution to acceptable levels. A(1) installations are generally perceived as having the greatest potential to pollute the environment. A(2) installations and Part B installations would have the least potential to pollute.

The industries affected by these regulations are those dealing with petrol vapour recovery, incineration of waste, mercury emissions from crematoria, animal rendering, non-ferrous foundry processes, surface treating of metals and plastic materials by powder coating, galvanizing of metals and the manufacture of certain specified composite wood-based boards.

Clean Air Act 1993

We are all entitled to breathe clean air but until quite recently the only method of heating houses and workshops was by burning coal, wood or peat in open fires. The smoke from these fires created air pollution and the atmosphere in large towns and cities was of poor quality. On many occasions in the 1950s the burning of coal in London was banned because the city was grinding to a halt because of the combined effect of smoke and fog, called smog. Smog was a very dense fog in which you could barely see more than a metre in front of you and which created serious breathing difficulties. In the new millennium we are no longer dependent upon coal and wood to heat our buildings. Smokeless coal has been created and the gaseous products of combustion are now diluted and dispersed by new chimney design regulations. Using well engineered combustion equipment together with the efficient arrestment of small particles in commercial chimneys of sufficient height, air pollution has been much reduced. This is what the Clean Air Act set out to achieve and it has been largely successful.

The **Clean Air Act** applies to all small and medium sized companies operating furnaces, boilers or incinerators. Compliance with the Act does not require an application for authorization and so companies must make sure that they do not commit an offence. In general the emission of dark smoke from any chimney is unacceptable. The emission of dark smoke from any industrial premises is also unacceptable. This might be caused by, for example, the burning of old tyres or old cable.

In England, Scotland and Wales it is not necessary for the Local Authority to have witnessed the emission of dark smoke before taking legal action. Simply the evidence of burned materials, which

Key fact

The *Clean Air Act* applies to all small and medium sized companies operating furnaces, boilers or incinerators.

potentially give rise to dark smoke when burned, is sufficient. In this way the law aims to stop people creating dark smoke under the cover of darkness.

A **public nuisance** is 'an act unwarranted by law or an omission to discharge a legal duty which materially affects the life, health, property, morals or reasonable comfort or convenience of Her Majesty's subjects'. This is a criminal offence and Local Authorities can prosecute, defend or appear in proceedings that affect the inhabitants of their area.

Controlled Waste Regulations 1992

Under these regulations we have a 'duty of care to handle, recover and dispose of all waste responsibly'. This means that all waste must be handled, recovered and disposed of by individuals or businesses that are authorized to do so under a system of signed Waste Transfer Notes.

The Environmental Protection (Duty of Care) Regulations 1991 state that as a business you have a duty to ensure that any waste you produce is handled safely and in accordance with the law. This is the 'duty of care' and applies to anyone who produces, keeps, carries, treats or disposes of waste from business or industry.

You are responsible for the waste that you produce, even after you have passed it on to another party such as a skip hire company, a scrap metal merchant, recycling company or local council. The duty of care has no time limit and extends until the waste has either been finally and properly disposed of or fully recovered.

So what does this mean for your company?

- Make sure that waste is only transferred to an authorized company.
- Make sure that waste being transferred is accompanied by the appropriate paperwork showing what was taken, where it was to be taken and by whom.
- Segregate the different types of waste that your work creates.
- Label waste skips and waste containers so that it is clear to everyone what type of waste goes into that skip.
- Minimize the waste that you produce and do not leave waste behind for someone else to clear away. Remember there is no time limit on your duty of care for waste.

Occupiers of domestic properties are exempt from the duty of care for the household waste that they produce. However, they do have a duty of care for the waste produced by, for example, a tradesperson working at a domestic property.

Key fact

Controlled Waste Regulations 1998
Under these regulations we have a 'duty of care to handle, recover and dispose of all waste responsibly'.

Special waste is covered by the Special Waste Regulations 1996 and is waste that is potentially hazardous or dangerous and which may, therefore, require special precautions during handling, storage, treatment or disposal. Examples of special waste are asbestos, lead-acid batteries, used engine oil, solvent-based paint, solvents, chemical waste and pesticides. The disposal of special waste must be carried out by a competent person, with special equipment and a licence.

Hazardous Waste Regulations 2005

The Hazardous Waste Regulations were introduced in July 2005 and under these regulations electric discharge lamps and tubes such as fluorescent, sodium, metal halide and mercury vapour are classified as hazardous waste. While each lamp only contains a very small amount of mercury, vast numbers are used and disposed of each year, resulting in a significant environmental threat. The environmentally responsible way to dispose of lamps and tubes is to recycle them and this process is now available through the electrical wholesalers.

Electrotechnical companies produce relatively small amounts of waste and even smaller amounts of special waste. Most companies buy in the expertise of specialist waste companies these days and build these costs into the contract.

Packaging (Essential Requirements) Regulations 2003

The Packaging Regulations were introduced on 25 August 2003 bringing the UK into harmony with Europe. The regulations deal with the essential requirements of packaging for the storage and transportation of goods. There are two essential elements to the regulations:

1 the packaging shall be designed and manufactured so that the volume and weight is to the minimum amount required to maintain the necessary level of safety for the packaged product and

2 the packaging shall be designed and manufactured in such a way that the packaging used is either reusable or recyclable.

The regulations are enforced by the Weights and Measures Authority in Great Britain, the Department of Enterprise, Trade and Investment in Northern Ireland and the Procurator-Fiscal or Lord Advocate in Scotland.

Waste Electrical and Electronic Equipment EU Directive 2007

The Waste Electrical and Electronic Equipment (WEEE) Regulations will ensure that Britain complies with its EU obligation to recycle waste from electrical products. The regulations came into effect in July 2007 and from that date any company which makes, distributes or trades in electrical or electronic goods such as household appliances, sports equipment and even torches and toothbrushes, will have to make arrangements for recycling these goods at the end of their useful life. Batteries are covered separately by another EU directive.

Some sectors are better prepared for the new regulations than others. Mobile phone operators, O2, Orange, Virgin and Vodafone, along with retailers such as Currys and Dixons, have already joined together to recycle their mobile phones collectively. In Holland the price of a new car now includes a charge for the recycling costs.

Further Information is available on the DTI and DEFRA websites under WEEE.

Key fact

The Waste Electrical and Electronic Equipment (WEEE) Regulations will ensure that Britain complies with its EU obligation to recycle waste from electrical products.

Radioactive Substances Act 1993

These regulations apply to the very low ionizing radiation sources used by specialized industrial contractors. The radioactive source may be sealed or unsealed. Unsealed sources are added to a liquid in order to trace the direction or rate of flow of that liquid. Sealed radioactive sources are used in radiography for the non-destructive testing of materials or in liquid level and density gauges.

This type of work is subject to the Ionising Radiations Regulations 1999 (IRR), which impose comprehensive duties on employers to protect people at work against exposure to ionizing radiation. These regulations are enforced by the HSE, while the Radioactive Substances Act (RSA) is enforced by the Environment Agency.

The RSA regulates the keeping, use, accumulation and disposal of radioactive waste, while the IRR regulates the working and storage conditions when using radioactive sources. The requirements of RSA are in addition to, and separate from, IRR for any industry using radioactive sources. These regulations also apply to offshore installations and to work in connection with pipelines.

Dangerous Substances and Preparations and Chemicals Regulations 2000

Chemical substances that are classified as carcinogenic, mutagenic or toxic, or preparations which contain those substances, constitute a risk to the general public because they may cause cancer, genetic disorders and birth defects, respectively.

These regulations were introduced to prohibit the supply of these dangerous drugs to the general public, to protect consumers from contracting fatal diseases through their use.

The regulations require that new labels be attached to the containers of these drugs which identify the potential dangers and indicate that they are restricted to professional users only.

The regulations implement Commission Directive 99/43/EC, known as the 17th Amendment, which brings the whole of Europe to an agreement that these drugs must not be sold to the general public, this being the only way of offering the highest level of protection for consumers.

The regulations will be enforced by the Local Authority Trading Standards department.

Noise regulations

Before 1960 noise nuisance could only be dealt with by common law as a breach of the peace under various Acts or local by-laws. In contrast, today there are many statutes, government circulars, British Standards and EU Directives dealing with noise matters. Environmental noise problems have been around for many years. During the eighteenth century, in the vicinity of some London hospitals, straw was put on the roads to deaden the sound of horses' hooves and the wheels of carriages. Today we have come a long way from this self-regulatory situation.

In the context of the Environmental Protection Act 1990, noise or vibration is a **statutory nuisance** if it is prejudicial to health or is a nuisance. However, nuisance is not defined and has exercised the minds of lawyers, magistrates and judges since the concept of nuisance was first introduced in the 1936 Public Health Act. There is a wealth of case law but a good working definition might be 'A statutory nuisance must materially interfere with the enjoyment of one's dwelling. It is more than just irritating or annoying and does not take account of the undue sensitivity of the receiver'.

The line that separates nuisance from no nuisance is very fine and non-specific. Next door's intruder alarm going off at 3 a.m. for an hour or more is clearly a statutory nuisance, whereas one going off a long way from your home would not be a nuisance. Similarly, an all night party with speakers in the garden would be a nuisance, whereas an occasional party finishing at say midnight would not be a statutory nuisance.

At Stafford Crown Court on 1 November 2004, Alton Towers, one of the country's most popular theme parks, was ordered by a judge to reduce noise levels from its 'white knuckle' rides. In the first judgment of its kind, the judge told the park's owners that neighbouring residents must not be interrupted by noise from rides such as Nemesis, Air, Corkscrew, Oblivion or from loudspeakers or fireworks.

The owners of Alton Towers, Tussauds Theme Parks Ltd, were fined the maximum sum of £5000 and served with a Noise Abatement Order for being guilty of breaching the 1990 Environmental Protection Act. Mr Richard Buxton, for the prosecution, said that the £5000 fine reflected the judge's view that Alton Towers had made little or no effort to reduce the noise nuisance.

Many nuisance complaints under the Act are domestic and are difficult to assess and investigate. Barking dogs, stereos turned up too loud, washing machines running at night to use 'low cost' electricity, television, DIY activities are all difficult to assess precisely as statutory nuisance. Similarly, sources of commercial noise complaints are also varied and include deliveries of goods during the night, general factory noises, refrigeration units, noise from public houses and clubs are all common complaints.

Industrial noise can be complex and complaints difficult to resolve both legally and technically. Industrial noise assessment is aided by BS 4142 but no guidance exists for other noise nuisance. The Local Authority has a duty to take reasonable steps to investigate all complaints and to take appropriate action.

The Noise and Statutory Nuisance Act 1993

This Act extended the statutory nuisance provisions of the Environmental Protection Act 1990 to cover noise from vehicles, machinery or equipment in the streets. The definition of equipment includes musical instruments but the most common use of this power is to deal with car alarms and house intruder alarms being activated for no apparent reason and which then continue to cause a nuisance for more than one hour.

In the case of a car alarm, a notice is fixed to the vehicle and an officer from the Local Authority spends one hour trying to trace the owner with help from the police and their national computer system. If the alarm is still sounding at the end of this period, then the Local Authority officer can break into the vehicle and silence the alarm. The vehicle must be left as secure as possible but if this cannot be done then it can be removed to a safe compound after the police have been notified. Costs can be recovered from the registered keeper.

Home intruder alarms that have been sounding for one hour can result in a 'Notice' being served on the occupier of the property, even if he or she is absent from the property at the time of the offence.

The Notice can be served by putting it through a letterbox. A Local Authority officer can then immediately silence the alarm without going into the property. However, these powers are *adoptive* and some Local Authorities have indicated that they will not adopt them because Sections 7–9 of the Act makes provision for incorporating the 'Code of Practice relating to Audible Intruder Alarms' into the statute. The two key points of the code are the installation of a 20 minute cut-off of the external sounder and the notification to the police and Local Authority of two key holders who can silence the alarm.

Noise Act 1996

This Act clarifies the powers which may be taken against work which is in default under the nuisance provisions of the Environmental Protection Act 1990. It provides a mechanism for permanent deprivation, return of seized equipment and charges for storage.

The Act also includes an *adoptive* provision making night time noise between 23:00 and 07:00 hours a criminal offence if the noise exceeds a certain level to be prescribed by the Secretary of State. If a notice is not complied with, a fixed penalty may be paid instead of going to court.

Noise at Work Regulations 1989

The Noise at Work Regulations, unlike the previous vague or limited provisions, apply to all work places and require employers to carry out assessments of the noise levels within their premises and to take appropriate action where necessary. The regulations came into force on 1 January 1990 implementing in the United Kingdom the EC Directive 86/188/EEC 'The Protection of Workers from Noise'.

Three action levels are defined by the regulations:

1 The first action level is a daily personal noise exposure of 85 dB, expressed as 85 dB(A).
2 The second action level is a daily personal noise exposure of 90 dB(A).
3 The third defined level is a peak action level of 140 dB(A) or 200 Pa of pressure which is likely to be linked to the use of cartridge operated tools, shooting guns or similar loud explosive noises. This action level is likely to be most important where workers are subjected to a small number of loud impulses during an otherwise quiet day.

The Noise at Work Regulations are intended to reduce hearing damage caused by loud noise. So, what is a loud noise? If you

> **Key fact**
>
> The Noise at Work Regulations are intended to reduce hearing damage caused by loud noise.

cannot hear what someone is saying when they are 2 m away from you or if they have to shout to make themselves heard, then the noise level is probably above 85 dB and should be measured by a competent person.

At the first action level an employee must be provided with ear protection (ear muffs or ear plugs) on request. At the second action level the employer must reduce, so far as is reasonably practicable, other than by providing ear protection, the exposure to noise of that employee.

Hearing damage is cumulative; it builds up, leading eventually to a loss of hearing ability. Young people, in particular, should get into the routine of avoiding noise exposure before their hearing is permanently damaged. The damage can also take the form of permanent tinnitus (ringing noise in the ears) and an inability to distinguish words of similar sound such as bit and tip.

Vibration is also associated with noise. Direct vibration, through vibrating floors or from vibrating tools, can lead to damage to the bones of the feet or hands. A condition known as 'vibration white finger' is caused by an impaired blood supply to the fingers, associated with vibrating hand tools.

Employers and employees should not rely too heavily on ear protectors. In practice they reduce noise exposure far less than is often claimed because they may be uncomfortable or inconvenient to wear. To be effective, ear protectors need to be worn all the time when in noisy places. If left off for even a short time, the best protectors cannot reduce noise exposure effectively.

Protection against noise is best achieved by controlling it at source. Wearing ear protection must be a last resort. Employers should:

- Design machinery and processes to reduce noise and vibration (mounting machines on shock absorbing materials can dampen out vibration).
- When buying new equipment, where possible choose quiet machines. Ask the supplier to specify noise levels at the operator's working position.
- Enclose noisy machines in sound absorbing panels.
- Fit silencers on exhaust systems.
- Install motor drives in a separate room away from the operator.
- Inform workers of the noise hazard and get them to wear ear protection.
- Reduce a worker's exposure to noise by job rotation or provide a noise refuge.

New regulations introduced in 2006 reduce the first action level to 80 dB(A) and the second level to 85 dB(A) with a peak action level of

98 dB(A) or 140 Pa of pressure. Every employer must make a 'noise' assessment and provide workers with information about the risks to hearing if the noise level approaches the first action level. He must do all that is reasonably practicable to control the noise exposure of his employees and clearly mark ear protection zones. Employees must wear personal ear protection whilst in such a zone.

The EHO (Environmental Health Officer)

The responsibilities of the EHO are concerned with reducing risks and eliminating the dangers to human health associated with the living and working environment. They are responsible for monitoring and ensuring the maintenance of standards of environmental and public health, including food safety, workplace health and safety, housing, noise, odour, industrial waste, pollution control and communicable diseases in accordance with the law. Although they have statutory powers with which to enforce the relevant regulations, the majority of their work involves advising and educating in order to implement public health policies.

The majority of EHO are employed by Local Authorities, who are the agencies concerned with the protection of public health. Increasingly, however, officers are being employed by the private sector, particularly those concerned with food, such as large hotel chains, airlines and shipping companies.

Your Local Authority EHO would typically have the responsibility of enforcing the environmental laws discussed above. Their typical work activities are to:

- ensure compliance with the Health and Safety at Work Act 1974, the Food Safety Act 1990 and the Environmental Protection Act 1990;
- carry out health and safety investigations, food hygiene inspections and food standards inspections;
- investigate public health complaints such as illegal dumping of rubbish, noise complaints and inspect contaminated land;
- investigate complaints from employees about their workplace and carry out accident investigations;
- investigate food poisoning outbreaks;
- obtain food samples for analysis where food is manufactured, processed or sold;
- visit housing and factory accommodation to deal with specific incidents such as vermin infestation and blocked drains;
- test recreational water, such as swimming pool water, and private water supplies in rural areas;
- inspect and licence pet shops, animal boarding kennels, riding stables and zoos;

Key fact

The responsibilities of the EHO are concerned with reducing risks and eliminating the dangers to human health associated with the living and working environment.

- monitor air pollution in heavy traffic areas and remove abandoned vehicles;
- work in both an advisory capacity and as enforcers of the law, educating managers of premises on issues which affect the safety of staff and members of the public.

In carrying out these duties, officers have the right to enter any workplace without giving notice, although notice may be given if they think it appropriate. They may also talk to employees, take photographs and samples and serve an Improvement Notice, detailing the work which must be carried out if they feel that there is a risk to health and safety that needs to be dealt with.

Enforcement law inspectors

If the laws relating to work, the environment and people are to be effective, they must be able to be enforced. The system of control under the Health and Safety at Work Act comes from the HSE or the Local Authority. Local Authorities are responsible for retail and service outlets such as shops, garages, offices, hotels, public houses and clubs. The HSE is responsible for all other work premises including the Local Authorities themselves. Both groups of inspectors have the same powers. They are allowed to:

- enter premises, accompanied by a police officer if necessary;
- examine, investigate and require the premises to be left undisturbed;
- take samples and photographs as necessary, dismantle and remove equipment;
- require the production of books or documents and information;
- seize, destroy or render harmless any substance or article;
- issue enforcement notices and initiate prosecutions.

There are two types of enforcement notices, an **'improvement notice'** and a **'prohibition notice'**.

An improvement notice identifies a contravention of the law and specifies a date by which the situation is to be put right. An appeal may be made to an Employment Tribunal within 21 days.

A prohibition notice is used to stop an activity which the inspector feels may lead to serious injury. The notice will identify which legal requirement is being contravened and the notice takes effect as soon as it is issued. An appeal may be made to the Employment Tribunal but the notice remains in place and work is stopped during the appeal process.

Cases may be heard in the Magistrates' or Crown Courts.

Key fact

If the laws relating to work, the environment and people are to be effective, they must be able to be enforced.

Key fact

An *improvement notice* identifies a contravention of the law and specifies a date by which the situation is to be put right.

Key fact

A *prohibition notice* is used to stop an activity which the inspector feels may lead to serious injury.

Magistrates' Court (Summary Offences) for health and safety offences, employers may be fined up to £20 000 and employees or individuals up to £5000. For failure to comply with an enforcement notice or a court order, anyone may be imprisoned for up to 6 months.

Crown Court (Indictable Offences) for failure to comply with an enforcement notice or a court order, fines are unlimited in the Crown Court and may result in imprisonment for up to 2 years.

Actions available to an inspector upon inspection of premises:

- Take no action – the law is being upheld.
- Give verbal advice – minor contraventions of the law identified.
- Give written advice – omissions have been identified and a follow up visit will be required to ensure that they have been corrected.
- Serve an improvement notice – a contravention of the law has taken, or is taking, place and the situation must be remedied by a given date. A follow up visit will be required to ensure that the matter has been corrected.
- Serve a prohibition notice – an activity has been identified which may lead to serious injury. The law has been broken and the activity must stop immediately;
- Prosecute – the law has been broken and the employer prosecuted.

On any visit, one or more of the above actions may be taken by the inspector.

The HSE and the EHO are the health and safety professionals. The day that one of these inspectors arrives to look at the health and safety systems and procedures that your company has in place is a scary day!

Environmental technology systems and renewable energy

Energy is vital to the modern industrial economy in the UK and Europe. We also need energy in almost every aspect of our lives, to heat and light our homes and offices, to enable us to travel on business or for pleasure, and to power our business and industrial machines.

In the past the UK has benefited from its fossil fuel resources of coal, oil and gas but respectable scientific sources indicate that the fossil fuel era is drawing to a close. Popular estimates suggest that gas and oil will reach peak production in the year 2060 with British coal reserves lasting only a little longer. Therefore we must look to different ways of generating electricity so that:

- the remaining fossil fuel is conserved
- our CO_2 emissions are reduced to avoid the consequences of climate change

- we ensure that our energy supplies are secure, and not dependent upon supplies from other countries.

Following the introduction of the Climate Change Act in 2008, the UK and other member states agreed an EU wide target of 20% renewable energy by the year 2020 and 60% by 2050. Meeting these targets will mean basing much of the new energy infrastructure around renewable energy, particularly offshore wind power.

The 'Energy Hierarchy' states that organizations and individuals should address energy issues in the following order so as to achieve the agreed targets:

1 Reduce the need for energy – reducing energy demand is cost saving, reduces greenhouse gas emissions and contributes to the security of supply. Reducing the energy loss from buildings by better insulation and switching off equipment when not in use is one way of achieving this target.
2 Use energy more efficiently – use energy efficient lamps and 'A' rated equipment.
3 Use renewable energy – renewable energy refers to the energy that occurs naturally and repeatedly in the environment. This energy may come from wind, waves or water, the sun or heat from the ground or air.
4 Any continuing use of fossil fuels should use clean and efficient technology. Power stations generating electricity from coal and oil (fossil fuel) release a lot of CO_2 in the generating process. New build power stations must now be fitted with carbon capture filters to reduce the bad environmental effects.

Funding for environmental technology systems

Renewable energy is no less reliable than energy generated from more traditional sources. Using renewable energy does not mean that you have to change your lifestyle or your domestic appliances. Being green is not about going back to the Middle Ages. Technology is part of the solution, not just the problem. There has never been a better time to consider generating energy from renewable technology than now because grants and funding are available to help individuals and companies.

The Low Carbon Building Programme, implemented by the Department of Energy and Climate Change (DECC), provides grants toward the installation of renewable technologies and is available to householders, public non-profit making organizations and commercial organizations across the UK.

Key fact

Renewable energy is no less reliable than energy generated from more traditional sources.

Key fact

The government's 'Feed in Tariff' pays a tax free sum which is guaranteed for 25 years.

The government's 'Feed in Tariff' pays a tax free sum which is guaranteed for 25 years. It is called 'clean energy cash back' and has been introduced to promote the uptake of small scale renewable and low carbon electricity generation technologies. The customer receives a generation tariff from the electricity supplier, whether or not any electricity generated is exported to the national grid, and an additional export tariff when electricity is transported to the electricity grid through a smart meter.

From April 2010, clean energy generators will be paid 41.3p for each kWh of electricity generated. Surplus energy fed back into the national grid earns an extra 3p per unit. If you add to this the electricity bill savings, a normal householder could be £1000 per year better off. Savings vary according to energy use and type of system used. The Energy Saving Trust at www.energysavingtrust.org.uk (telephone number 01752 823600) and www.britishgas.co.uk and Ofgem at www.ofgem.gov.uk/fits provide an on line calculator to determine the cost, size of system and CO_2 savings for PV systems.

Micro-generation technologies

Today, micro-generation systems generate relatively small amounts of energy at the site of a domestic or commercial building. However, it is estimated that by 2050, 30 to 40% of the UK's electricity demand could be met by installing micro-generation equipment to all types of building.

Key fact

Micro-generation technologies include small wind turbines, solar photovoltaic (PV) systems, small scale hydro and micro CHP (combined heat and power) systems.

In the USA, the EU and the UK, buildings consume more than 70% of the nation's electricity and contribute almost 40% of the polluting CO_2 greenhouse gases. Any reductions which can be made to these figures will be good for the planet, and hence the great interest today in micro-generation systems. Micro-generation technologies include small wind turbines, solar photovoltaic (PV) systems, small scale hydro and micro CHP (combined heat and power) systems. Micro-generators that produce electricity may be used as stand alone systems, or may be run in parallel with the low voltage distribution network, that is, the A.C. mains supply.

The April 2008 amendments to the Town and Country Planning Act 1990 now allow the installation of micro-generation systems within the boundary of domestic premises without obtaining planning permission. However, size limitations have been set to reduce the impact upon neighbours. For example, solar panels attached to a building must not protrude more than 200 mm from the roof slope and stand alone panels must be no higher than four metres above ground level and no nearer than five metres from the property boundary. See the Electrical Safety Council site for advice on connecting micro-generation systems at www.esc.org.uk/bestpracticeguides.html

Smart electricity meters

Smart electricity meters are designed to be used in conjunction with micro-generators. Electricity generated by the consumer's micro-generator can be sold back to the energy supplier using the 'smart' two way meter.

The Department for Energy and Climate Change is planning to introduce smart meters into consumers' homes from 2012 and this is expected to run through until 2020 with the aim being to help consumers reduce their energy bills.

When introducing the proposal, Ed Miliband, the then Energy and Climate Change Secretary, said "the meters which most of us have in our homes were designed for a different age, before climate change. Now we need to get smarter with our energy. This is a big project affecting 26 million homes and several million businesses. The project will lead to extra work for electrical contractors through installing the meters on behalf of the utility companies and implementing more energy efficient devices once customers can see how much energy they are using".

Already available now is the Real Time Display (RTD) wireless monitor which enables consumers to see exactly how many units of electricity they are using through an easy to read portable display unit. By seeing the immediate impact in pence per hour of replacing existing lamps with low energy ones or switching off unnecessary devices throughout the home or office, consumers are naturally motivated to consider saving energy. RTD monitors use a clip on sensor on the meter tails and include desk top software for PC and USB links.

Let us now look at some of these micro-generation technologies.

> **Key fact**
>
> Smart electricity meters are designed to be used in conjunction with micro-generators.

Solar photovoltaic (PV) power supply systems

A solar photovoltaic (PV) system is a collection of PV cells known as a PV string, that forms a PV array and collectively are called a PV generator, which turns sunlight directly into electricity. PV systems may be 'stand alone' power supplies or be designed to operate in parallel with the public low voltage distribution network, that is the A.C. mains supply.

Stand alone PV systems are typically a small PV panel of maybe 300 mm by 300 mm tilted to face the southern sky, where it receives the maximum amount of sunlight. They typically generate 12 to 15 volts and are used to charge battery supplies on boats, weather stations, road signs and any situation where electronic equipment is used in remote areas away from a reliable electrical supply.

> **Key fact**
>
> A solar photovoltaic (PV) system turns sunlight directly into electricity.

Figure 5.1 Photograph of PV system in a domestic situation.

Developing nations are beginning to see stand alone PV systems as the way forward for electrification of rural areas rather than continuing with expensive diesel generators and polluting kerosene lamps.

The cost of PV generators is falling. The period 2009 to 2010 saw the cost of PV cells fall by 30% and with new 'thin-film' cells being developed, the cost is expected to continue downwards. In the rural areas of the developing nations they see PV systems linked to batteries bringing information technology, radio and television to community schools. This will give knowledge and information to the next generation which will help these countries to develop a better economy, a better way of life and to have a voice in the developed world.

Stand alone systems are not connected to the electricity supply system and are therefore exempt from much of BS 7671, the IEE Regulations. However IEE Regulation 134.1.1 'good workmanship by competent persons and proper materials shall be used in all electrical installations' will apply to any work done by an electrician who must also pay careful attention to the manufacturer's installation instructions.

Solar photovoltaic (PV) systems designed to operate in parallel with the public low voltage distribution network are the type of micro-generator used on commercial and domestic buildings. The PV cells operate in exactly the same way as the stand alone system described above, but will cover a much greater area. The PV cells are available in square panels which are clipped together and laid over the existing roof tiles as shown in Fig. 5.1 or the PV cells may be manufactured to look just like the existing roof tiles which are integrated into the existing roof.

A solar PV system for a domestic three bedroom house will require approximately 15 to 20 m^2 generating two to three kilowatts of power and the cost of the PV cells alone will be in the region of £10 000 to £12 000, although grants are available. On the positive side, a PV system for a three bedroom house will save around 1200 kg of CO_2 per year.

These bigger micro-generation systems are designed to be connected to the power supply system and the installation must therefore comply with Section 712 of BS 7671:2008. Section 712 contains the requirements for protective measures comprising automatic disconnection of the supply wiring systems, isolation, switching and control, earthing arrangements and labelling. In addition, the installation must meet the requirements of the Electricity Safety Quality and Continuity Regulations 2006. This is a mandatory requirement. However, where the output does not exceed 16 A per line, they are exempt from some of the requirements providing that:

- the equipment is type tested and approved by a recognized body
- the installation complies with the requirements of BS 7671, the IEE Regulations
- the PV equipment must disconnect from the distributor's network in the event of a network fault
- the distributor must be advised of the installation before or at the time of commissioning

Installations of less than 16 A per phase but up to 5 kilowatt peak (kWp) will also be required to meet the requirements of the Energy Network Association's Engineering Recommendation G83/1 for small scale embedded generators in parallel with public low voltage distribution networks. Installations generating more than 16 A must meet the requirements of G59/1 which requires approval from the distributor before any work commences.

Solar thermal hot water heating

Solar thermal hot water heating systems are recognized as a reliable way to use the energy of the sun to heat water. The technology is straightforward and solar thermal panels for a three bedroom house cost between £3000 and £6000 for a 3 to 6 m^2 panel and they will save about 260 kg of CO_2 annually.

The solar panel comprises a series of tubes containing water that is pumped around the panel and a heat exchanger in the domestic water cylinder as shown in Fig. 5.2. Solar energy heats up the

Key fact

Solar thermal hot water heating systems are recognized as a reliable way to use the energy of the sun to heat water.

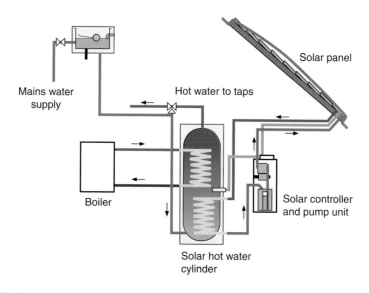

Mains water supply

Hot water to taps

Solar panel

Boiler

Solar hot water cylinder

Solar controller and pump unit

Figure 5.2 Solar powered hot water system.

domestic hot water. A solar panel of about 4 m² will deliver about 1000 kWh per year which is about half the annual water demand of a domestic dwelling. However, most of the heat energy is generated during the summer and so it is necessary to supplement the solar system with a boiler in the winter months. Figure 5.3 shows a photograph of an installed solar hot water panel.

If you travel to Germany, you will see a lot of photovoltaic and solar thermal panels on the roofs there. In the UK, planning rules for solar thermal and PV installations have already been made much easier. A website detailing planning requirements for solar and wind can be found at www.planningportal.gov.uk/uploads/hhghouseguide.html

Wind energy generation

Modern large scale wind machines are very different from the traditional windmill of the last century which gave no more power than a small car engine. Very large structures are needed to extract worthwhile amounts of energy from the wind. Modern large scale wind generators are taller than electricity pylons, with a three blade aeroplane-type propeller to catch the wind and turn the generator. If a wind turbine was laid down on the ground, it would be longer and wider than a football pitch. They are usually sited together in groups in what have become known as 'wind energy farms' as shown in Fig. 5.4.

Each modern grid-connected wind turbine generates about 600 kW of electricity. A wind energy farm of 20 generators will therefore generate 12 MW, a useful contribution to the national grid, using a

Figure 5.3 Photograph of a solar hot water panels.

Figure 5.4 Wind energy farm generators.

naturally occurring, renewable, non-polluting source of energy. The Department of Energy and Climate Change considers wind energy to be the most promising of the renewable energy sources.

In 2010 there were 253 wind energy farms in operation in the UK with 12 operating offshore. The 3000 turbines on these farms have the capacity to generate 4600 MW of electricity, enough for 2.5 million homes. There are a further 500 wind energy farms planned or in construction. However, because of the unpredictable nature of the wind, and inefficiencies in the generation process, the amount of power produced will be less than the installed capacity.

Countryside Commission, the government's adviser on land use, has calculated that to achieve a target of generating 10% of the total electricity supply by wind power will require 40 000 generators of the present size. At the time of writing, 2010, we are generating only about 2.5% of the total electricity supply from wind power and all hopes are pinned on large offshore wind farms to achieve the government and EU targets for 2020.

Wind power is an endless renewable source of energy, is safer than nuclear power and provides none of the polluting emissions associated with fossil fuel. If there was such a thing as a morally pure form of energy, then wind energy would be it. However, wind farms are, by necessity, sited in some of the most beautiful landscapes in the UK. Building wind energy farms in these areas of outstanding natural beauty has outraged conservationists. Prince Charles has reluctantly joined the debate saying that he was in favour of renewable energy sources but believed that 'wind farms are an horrendous blot on the landscape'. He believes that if they are to be built at all they should be constructed well out to sea.

The next generation of wind farms will mostly be built offshore, and this is the place to build them, where there is more space and more wind, but the proposed size of these turbines creates some big engineering problems. From the sea bed foundations to the top of the turbine blade will be up to a staggering 250 m, three times the height of the Statue of Liberty. Each offshore turbine, generating between 5 and 7 MW, will weigh between 200 and 300 tonnes. When you put large wind forces onto that structure you will create a very big cantilever effect which creates a very big engineering challenge.

The world's largest offshore wind energy farm built so far was opened in September 2010. The 100 turbine 'Thanet' wind farm just off the Kent coast will generate enough power to supply 200 000 homes. The Thanet project cost £780 million pounds to build. The turbines are up to 380 feet high and cover an area as large as 4000 football pitches. The Thanet project did not retain its title as the world's largest wind farm for long because the 'Greater Gabbard' wind farm, off the east

Key fact

Wind power is an endless renewable source of energy, is safer than nuclear power and provides none of the polluting emissions associated with fossil fuel.

Figure 5.5 Small wind generator on a domestic property.

coast and having 140 turbines, and opened in late 2010. The Thanet project brings Britain's total wind energy capacity above 5000 MW for the first time.

The Department of Energy and Climate Change has calculated that 10 000 wind turbines could provide the energy equivalent of 8 million tonnes of coal per year and reduce CO_2 emissions. While this is a worthwhile saving of fossil fuel, opponents point out the obvious disadvantages of wind machines, among them the need to maintain the energy supply during periods of calm which means that wind machines can only ever supplement a more conventional electricity supply.

Small wind micro-generators can be used to make a useful contribution to a domestic property or a commercial building. They can be stand alone and are about the size of a tall street lamp. A 12 m high turbine costs about £24 000 and, with a good wind, will generate 10 000 kWh per year. However, if you live in a village, town or city you are unlikely to obtain the Local Authority building and planning permissions to install a wind generator because your neighbours will object.

Small wind generators of the type shown in Fig. 5.5 typically generate between 1.5 A and 15 A in wind speeds of 10 mph to 40 mph.

Heat pumps

In applications where heat must be upgraded to a higher temperature so that it can be usefully employed, a heat pump must be used. Energy from a low temperature source such as the earth, the air, a lake or river is absorbed by a gas or liquid contained in pipes, which is then mechanically compressed by an electric pump to produce a temperature increase. The high temperature energy is then transferred to a heat exchanger so that it might do useful work, such as providing heat to a building. For every 1 kWh of electricity used to power the heat pump compressor, approximately 3 to 4 kWh of heating are produced.

How a heat pump works

1 A large quantity of low grade energy is absorbed from the environment and transferred to the refrigerant inside the heat pump (called the evaporator). This causes the refrigerant temperature to rise, causing it to change from liquid to a gas.
2 The refrigerant is then compressed, using an electrically driven compressor, reducing its volume but causing its temperature to rise significantly.
3 A heat exchanger (condenser) extracts the heat from the refrigerant to heat the domestic hot water or heating system.
4 After giving up its heat energy, the refrigerant turns back into a liquid and, after passing through an expansion valve, is once more ready to absorb energy from the environment and the cycle is repeated as shown in Fig 5.6.

A refrigerator works on this principle. Heat is taken out of the food cabinet, compressed and passed on to the heat exchanger or radiator at the back of the fridge. This warm air then radiates by air convection currents into the room. Thus the heat from inside the cabinet is moved into the room leaving the sealed refrigerator cabinet cold.

Heat pumps (ground source)

Ground source heat pumps extract heat from the ground by circulating a fluid through polythene pipes buried in the ground in trenches or in vertical boreholes as shown in Fig. 5.7. The fluid in the pipes extracts heat from the ground and a heat exchanger within the pump extracts heat from the fluid. These systems are most effectively used to provide underfloor radiant heating or water heating.

Calculations show that the length of pipe buried at a depth of 1.5 m required to produce 1.2 kW of heat will vary between 150 m in dry soil and 50 m in wet soil. The average heat output can be taken as 28

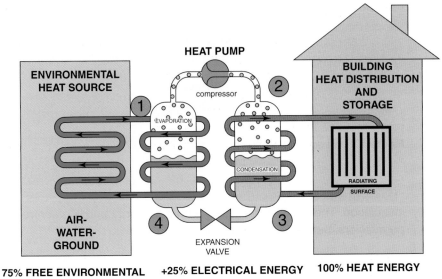

Figure 5.6 Heat pump working principle.

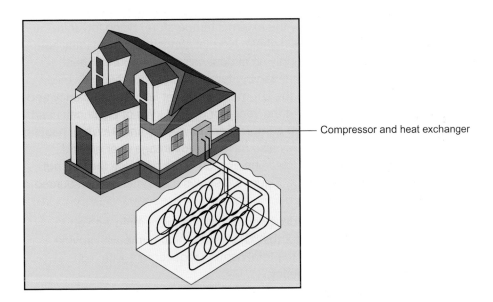

Figure 5.7 Ground source heating system.

watts per metre of pipe. A rule of thumb guideline is that the surface area required for the ground heat exchanger should be about 2.5 times the area of the building to be heated.

This type of installation is only suitable for a new build project and the ground heat exchanger will require considerable excavation and

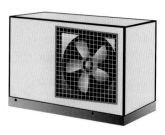

Figure 5.8 Air sourced heat pump unit.

installation. The installer must seek Local Authority building control permissions before work commences.

Heat pumps (air source)

The performance and economics of heat pumps are largely determined by the temperature of the heat source and so we seek to use a high temperature source. The heat sources used by heat pumps may be soil, the air, ground or surface water. Unfortunately all these sources follow the external temperature, being lower in winter when demand is highest. Normal atmosphere is an ideal heat source in that it can supply an almost unlimited amount of heat although unfortunately at varying temperatures, but relatively mild winter temperatures in the UK mean excellent levels of efficiency and performance throughout the year. For every 1 kWh of electricity used to power the heat pump compressor, between 3 and 4 kWh of heating energy is produced. They also have the advantage over ground source heat pumps of lower installation costs because they do not require any groundwork. Figure 5.8 shows a commercial air sourced heat pump.

If the air heat pump is designed to provide full heating with an outside temperature of 2 to 4°C, then the heat pump will provide approximately 80% of the total heating requirement with high performance and efficiency.

The point at which the output of a given heat pump meets the building heat demand is known as the 'balance point'. In the example described above, the 20% shortfall of heating capacity below the balance point must be provided by some supplementary heat. However, an air to air heat pump can also be operated in the reverse cycle which then acts as a cooling device, discharging cold air into the building during the summer months. So here we have a system which could be used for air conditioning in a commercial building.

Hydroelectric power generation

The UK is a small island surrounded by water. Surely we could harness some of the energy contained in tides, waves, lakes and rivers? Many different schemes have been considered in the past 20 years and a dozen or more experimental schemes are being tested now.

Water power makes a useful contribution to the energy needs of Scotland but the possibility of building similar hydroelectric schemes in England are unlikely chiefly due to the topographical nature of the country.

The Severn Estuary has a tidal range of 15 m, the largest in Europe, and a reasonable shape for building a dam across the estuary. This would allow the basin to fill with water as the tide rose, and then allow the impounded water to flow out through electricity generating turbines as the tide falls. However, such a tidal barrier might have disastrous ecological consequences upon the many wildfowl and wading bird species because of the submerging of the mudflats which now provide winter shelter for these birds. Therefore, the value of the power which might be produced must be balanced against the possible ecological consequences.

France has successfully operated a 240 MW tidal power station at Rance in Brittany for the past 25 years.

Marine Current Turbines Ltd are carrying out research and development on submerged turbines which will rotate by exploiting the principle of flowing water in general and tidal streams in particular. The general principle is that an 11 m diameter water turbine is lowered into the sea down a steel column drilled in the sea bed. The tidal movement of the water then rotates the turbine and generates electricity.

The prototype machines were submerged in the sea off Lynmouth in Devon. In May 2008 they installed the world's first tidal turbine in the Strangford Narrows in Northern Ireland where it is now grid connected and generating 1.2 MW.

All the above technologies are geared to providing hydroelectric power connected to the national grid, but other micro-hydro schemes are at the planning and development stage.

Micro hydro generation

The use of small hydro power (SHP) or micro hydro power has grown over recent decades led by continuous technical developments, brought about partly in the UK by the new Coalition Government's 'feed in tariff' where green electricity producers are paid a premium to produce electricity from renewable sources.

The normal perception of hydro power is of huge dams, but there is a much bigger use of hydro power in smaller installations. Asia, and especially China, is set to become a leader in hydroelectric generation. Australia and New Zealand are focusing on small hydro plants. Canada, a country with a long tradition of hydro power, is developing small hydro power as a replacement for expensive diesel generation in remote off-grid communities.

Small hydro power schemes generate electricity by converting the power available in rivers, canals and streams. The object of a hydro power scheme is to convert the potential energy of a mass of water flowing in a stream with a certain fall, called the head, into electrical energy at the lower end of the stream where the power house is located. The power generated is proportional to the flow, called the discharge, and to the head of water available. The fundamental asset of hydro power is that it is a clean and renewable energy source and the fuel used, water, is not consumed in the electricity generating process.

In the Derbyshire Peak District along the fast flowing River Goyt there were once 16 textile mills driven by water wheels. The last textile mill closed in 2000 but the old Torr Mill has been saved. Where once the water wheel stood is now a gigantic 12 tonne steel screw, 2.4 m in diameter. The water now drives the reverse Archimedean screw, affectionately called 'Archie', to produce enough electricity for 70 homes. The electricity generating project is owned by the residents of New Mills in a sharing co-operative in which surplus electricity is sold back to the grid. Archie will produce 250 MWh of electricity each year and the installation cost was £300 000 in 2008.

In July 2010 The Lake District National Park Authority granted permission for a water turbine to be built on a 350 m long stretch of the fast flowing River Kent. Approximately 1225 MWh of electricity will be generated each year, enough to power 250 homes. The project will cost £1.6 M funded by grants and loans and completion is anticipated in late 2011 or early 2012. The Kentmere Hydro Project is a community project and the power generated will be fed back to the national grid. The Trust will receive money from the government's clean energy cash back scheme which it will use to pay back the construction loans, leaving a surplus which will be used to support local projects in this small Lakeland community.

The type of turbine chosen for any hydro scheme will depend upon the discharge rate of the water and the head of water available.

Key fact

The normal perception of hydro power is of huge dams, but there is a much bigger use of hydro power in smaller installations.

Key fact

Small hydro power schemes generate electricity by converting the power available in rivers, canals and streams.

Figure 5.9 An example of an Archimedian screw at the River Dart country park, Devon. Credit: WRE Limited.

A Pelton wheel is a water turbine in which specially shaped buckets attached to the periphery of the wheel are struck by a jet of water. The kinetic energy of the water turns the wheel which is coupled to the generator.

Axial turbines comprise a large cylinder in which a propeller type water turbine is fixed at its centre. The water moving through the cylinder causes the turbine blade to rotate and generate electricity.

A Francis turbine or kaplan turbine is also an axial turbine but the pitch of the blades can be varied according to the load and discharge rate of the water.

Small water turbines will reach a mechanical efficiency at the coupling of 90%.

Up and down the country, riverside communities must be looking at the relics of our industrial past and wondering if they might provide a modern solution for clean, green, electrical energy. However, despite the many successes and and obvious potential, there are many barriers to using waterways for electricity generation in European countries. It is very difficult in this country to obtain permission from British Waterways to extract water from rivers, even though once the water has passed through the turbine, it is put back into the river. Environmental pressure groups are opposed to micro hydro generation because of its perceived local environmental impact on the river ecosystem and the disturbance to fishing. Therefore, once again, the value of the power produced would have to be balanced against the possible consequences.

Combined heat and power (CHP)

CHP is the simultaneous generation of usable heat and power in a single process. That is, heat is produced as a by-product of the power generation process. A chemical manufacturing company close to where I live has a small power station which meets some of their electricity requirements using the smart meter principle. Their 100 MW turbine is driven by high pressure steam. When the steam condenses after giving up its energy to the turbine, there remains a lot of very hot water which is then piped around the offices and some production plant buildings for space heating. Combining heat and power in this way can increase the overall efficiency of the fuel used because it is performing two operations.

CHP can also use the heat from incinerating refuse to heat a nearby hospital, school or block of flats.

> **Key fact**
>
> The type of turbine chosen for any hydro scheme will depend upon the discharge rate of the water and the head of water available.

> **Key fact**
>
> CHP is the simultaneous generation of usable heat and power in a single process. That is, heat is produced as a by-product of the power generation process.

Biomass heating

Biomass is derived from plant materials and animal waste. It can be used to generate heat and to produce fuel for transportation. The biomass material may be straw and crop residues, crops grown specially for energy production, rape seed oil and waste from a range of sources including food production. The nature of the fuel will determine the way that energy can best be recovered from it.

There is a great deal of scientific research being carried out at the moment into 'Biomass Renewables', that is energy from crops. This area of research is at an early stage, but is expected to flourish in the next decade. The first renewable energy plant, which is to be located at Teesport on the river Tees in the north east of England, has received approval from the Department for Energy and Climate Change for building to commence.

The facility will be one of the largest biomass plants to be built in the world and is scheduled to enter commercial operation in late 2012. Young trees will be grown as a crop to produce wood chips. The plant will use 2.5 million tonnes of wood chips each year to produce 300 MW of electrical energy. The plant will operate 24 hours a day, all year round to meet some of the national grid base load.

Water conservation

Conservation is the preservation of something important, especially of the natural environment. Available stored water is a scarce resource in England and Wales where there is only 1400 cubic meters per person per year. Very little compared with France, which has 3100 cubic meters per person per year, Italy which has 2900 and Spain 2800. About a half of the water used by an average home is used to shower, bathe and wash the laundry, another third is used to flush the toilet.

At a time when most domestic and commercial properties have water meters installed, it saves money to harvest and reuse water.

The City & Guilds have asked us to look at two methods of water conservation, by rainwater harvesting and grey water recycling.

Rainwater harvesting

Rainwater harvesting is the collection and storage of rainwater for future use. Rainwater has in the past been used for drinking, water for livestock and water for irrigation. It is now also being used to provide water for car cleaning and garden irrigation in domestic and commercial buildings.

Many gardeners already harvest rainwater for garden use by collecting run off from a roof or greenhouse and storing it in a water butt or water barrel. However a 200 litre water butt does not give much drought protection although garden plants much prefer rain water to fluoridated tap water. To make a useful contribution the rainwater storage tank should be between 2000 and 7000 litre capacity. The rainwater collecting surfaces will be the roof of the building and any hard paved surfaces such as patios. Down pipes and drainage pipes then route the water to the storage tank situated, perhaps, under the lawn. An electric pump lifts the water from the storage tank to the point of use, possibly a dedicated outdoor tap. The water is then distributed through a hose pipe or sprinkler system to the garden in the normal way.

With a little extra investment, rainwater can be filtered and used inside the house to supply washing machines and WCs. Installing domestic pipes and interior plumbing can be added to existing homes although it is more straightforward in a new build home.

With the move toward more sustainable homes, UK architects are becoming more likely to specify rainwater harvesting in their design to support alternatives to a mains water supply. In Germany rainwater harvesting systems are now installed as standard in all new commercial buildings.

Grey water recycling

Grey water is tap water which has already performed one operation and is then made available to be used again instead of flushing it down the drain. Grey water recycling offers a way of getting double the use out of the world's most precious resource.

There are many products on the market, such as the BRAC System which takes in water used in the shower, bath and laundry, cleans it by filtering and then reuses it for toilet flushing. It is only a matter of routing the grey waste water drain pipe from the bath, shower and laundry to the filter unit and then plumbing the sanitized grey water to the toilet tank.

These systems are easy to install, particularly in a new build property. It is only a matter of re-routing the drain pipes. Another option for your grey water is to route it into the rainwater storage tank for further use in the garden.

> **Key fact**
>
> Grey water is tap water which has already performed one operation and is then made available to be used again instead of flushing it down the drain.

Water Regulations 1999

Water supply and installations in England and Wales are controlled by the Water Regulations known as the Water Supply (Water Fittings)

Regulations 1999 which came into force on 1 July 1999. Separate arrangements apply to Scotland and Northern Ireland.

The water that finds its way to our taps is derived from rainfall and the treatment of that water depends upon where it is sourced from and the impurities it contains. Water sourced from springs and wells is naturally purified and needs little disinfection. The quality of the water from reservoirs and rivers, called raw water, will determine the level of treatment. This usually involves several stages of treatment including settling, filtering and a final 'polishing' with carbon grains to remove minute traces of impurities and to improve the water taste. Water suppliers store water either in its raw state in impounding reservoirs or lakes, or as treated wholesome water in service reservoirs.

After being treated, water is distributed from the water supplier to individual consumers through a network of pipes known as 'mains'. The mains belong to the water supplier and it is their responsibility to maintain them in a way that will conserve this important resource. The local mains provide the final leg of the journey to our homes.

The water at our taps is of the very highest quality and it seems a little irresponsible to flush it down the drains without giving some consideration to water conservation.

Energy saving legislation

In April 2006, Part L of the Building Regulations (England and Wales) was revised in order to raise energy performance standards and to reduce CO_2 emissions from buildings. Part L, Conservation of Fuel and Power now requires all new and existing buildings to be given an energy rating and for all new buildings to meet a minimum level of energy efficiency. Under this provision the electrical contractor must make 'reasonable provision' to provide lighting systems with energy efficient lamps and sufficient controls so that electrical energy can be used efficiently. The current provision requires one energy efficient luminaire for every $25\,m^2$ of floor area or one energy efficient luminaire for every four fixed luminaires.

External lighting fixed to the building, including lighting in porches, but not garages or car ports, should provide reasonable provision for energy efficient lamps such as fluorescent tubes or CFLs. These lamps should automatically extinguish in daylight and when not required at night, being controlled by passive infra-red (PIR) detectors.

The traditional carbon filament lamp, called a GLS (general lighting service) lamp, is hopelessly bad in energy efficient terms, producing only 14 lumens of light output for every electrical watt input.

Key fact

The water that finds its way to our taps is derived from rainfall and the treatment of that water depends upon where it is sourced from and the impurities it contains.

Key fact

In April 2006, Part L of the Building Regulations (England and Wales) was revised in order to raise energy performance standards and to reduce CO_2 emissions from buildings.

Fluorescent tubes and CFLs produce more than 40 lumens of light output for every electrical watt input.

In addition, the electrical installer must have an appreciation of how the building regulations in general might affect the electrical installation in particular. For example:

- Part A Structure – the basic requirement for those installing electrical installations in a building is not to cut, drill, chase, penetrate or in any way interfere with the structure so as to cause significant reduction in its load bearing capability. Approved document A provides practical guidance with pictures. This document can be found in the Electrician's Guide to the Building Regulations published by the IEE.
- Part B Fire Safety – the 'standard house' with no floor area exceeding $200\,m^2$ must be fitted with smoke alarms to each level. If the kitchen cannot be isolated from the other rooms by a door, then a compatible interlinked heat detector must also be installed in the kitchen.
- Part M Access and Facilities for the Disabled – this requires switches and socket outlets in dwellings to be installed so that all persons, including those whose reach is limited, can easily reach them. The recommendation is that they should be installed in habitable rooms at a height of between 450 and 1200 mm from the finished floor level as shown in Fig. 4.10 of chapter 4.

The important change in the 2006 regulations is that compliance is now based upon the the whole building's carbon emissions, meaning that the building designer must now consider the impact of both the constructional elements of the building as well as the energy using services within the building such as lighting, heating, hot water and ventilation.

To achieve compliance, the building designer must show that the predicted annual carbon emissions from the building are less than, or equal to, a Target Carbon Emissions Rate for a 'standard national building' of the same floor area and shape as the one being designed.

Code for sustainable homes

The use of energy to provide heat for central heating and hot water in our homes is responsible for 60% of a typical family's energy bill. Heating accounts for over half of Britain's entire use of energy and carbon emissions. If Britain is to reduce its carbon footprint and achieve energy security, we must revolutionize the way we keep warm in the home.

At present 69% of our home heating comes from burning gas, 11% from oil, 3% from solid fuels such as coal and 14% from electricity

Key fact

Heating accounts for over half of Britain's entire use of energy and carbon emissions.

Date	2010	2013	2016
Energy efficiency improvement of the dwelling compared to 2006 (part L Building Regulations)	2.5%	44%	Zero carbon
Equivalent Standard within the Code	Code level 3	Code level 4	Code level 6

Figure 5.10 Code for Sustainable Homes.

which is mainly generated from these same three fossil fuels. Only 1% is currently provided by renewable sources. If Britain is to meet its clean energy targets, renewable sources will have to increase, and the revolution will have to start in the home because the country's dwellings currently provide more than half of the total demand, almost entirely for hot water and central heating.

There are about 20 million homes in the UK and a review of present buildings has found that about six million homes have inadequately lagged lofts, eight million have uninsulated cavity walls and a further seven million homes with solid walls would benefit from better insulation. If the country is to achieve its 2016 reduced carbon emissions targets, these existing homes must be heavily insulated to reduce energy demand and then supplied with renewable heat.

We cannot sustain the present level of carbon emissions without disastrous ecological consequences in the future. Low carbon homes are sustainable homes.

HRH The Prince of Wales has entered the debate saying "becoming more sustainable is possibly the greatest challenge humanity has faced and I am convinced that it is, therefore, the most remarkable chance to secure a prosperous future for everyone. We must strive harder than ever before, to convince people that by living sustainably we will improve our quality of life and our health; that by living in harmony with nature we will protect the intricate, delicate balance of the natural systems that ultimately sustain us" (Daily Telegraph, 31 July 2010).

The Code for Sustainable Homes, see Fig. 5.10, measures the sustainability of a home against categories of sustainable design, rating

Key fact

The Code for Sustainable Homes, see Fig. 5.10, measures the sustainability of a home against categories of sustainable design, rating the whole home as a complete package.

the whole home as a complete package, including building materials and services within the building. The code uses a 1 to 6 star rating to communicate the overall sustainability performance of a new home and sets minimum standards for energy and water use at each level.

Since May 2008 all new homes are required to have a Code Rating and a Code Certificate. By 2016 all new homes must be built to zero carbon standards, which will be achieved through step by step tightening of the Building Regulations.

If we look at sustainability from a manufacturing point of view, sustainable manufacture is based on the principle of meeting the needs of the current generation without compromising the ability of future generations to meet their needs.

Safe working environment

Whatever type of work you are involved with in the electrotechnical industry, electrical installation work, maintenance or installing eco-friendly systems, you must always create a safe working environment not only for your own safety but for the safety of others who might be affected by your actions.

In Chapter 1 we looked at some of the laws and regulations that affect our working environment. We looked at safety signs and PPE and how to recognize and use different types of fire extinguishers. The structure of companies within the electrotechnical industry and the ways in which they communicate information by drawings, symbols and standard forms was also discussed.

We began to look at safe electrical isolation procedures in Chapter 1 and then discussed this topic further in Chapter 3. Safe manual handling techniques and safe procedures for working above ground level were shown in Figs 3.3 to 3.7.

In Chapter 3, under the heading 'Avoiding accidents in the workplace' we looked at the common causes of accidents at work and how to control the risks associated with various hazards. At Fig. 3.2 we looked at the 'Hazard risk assessment' process.

In Chapter 4 we looked at the installation of electrical equipment and wiring systems in buildings, including some of the 'special' installations.

In this chapter we have looked at some of the environmental laws and the new environmental technology systems for generating electricity. Compare these micro-generation systems with the power required for the national grid discussed in Chapter 2 and Fig. 2.16.

If your career in the electrotechnical industry is to be a long, happy and safe one, you must always behave responsibly and sensibly in

Key fact

When you have read and understood the whole of this chapter you have completed all the environmental legislation and new environmental technology systems outcomes of the City & Guilds 2357 knowledge units.

When you have completed the practical assessments required by the City & Guilds syllabus, which you are probably doing at your local college, you may be ready to tackle the online assessment. So, to prepare you for the online assessment, try the following assessment questions.

order to maintain a safe working environment. Before starting work, make a safety assessment - what is going to be hazardous – will you require PPE - do you need any special access equipment? Carry out safe isolation procedures before beginning any work. You do not necessarily have to do these things formally, such as carrying out the risk assessment described in Chapter 3, but just get into the habit of always working safely and being aware of the potential hazards around you when you are working.

Do not leave your tools lying around for others to fall over or steal. Keep them close by you in a toolbox. The tools and equipment which you are not using should be locked away in a safe storage place.

Finally, when a job is finished, clean up and dispose of all waste material responsibly as described at the end of chapter 4.

Try This

This might be a good time to revise some of the safety at work topics discussed in Chapter 3.

- *Read again 'Avoiding accidents in the workplace' on page 83*
- *Look at the 'Hazard risk assessment' process shown in Fig. 3.2*
- *Read again the section on 'Safe working above ground level on page 88*
- *And finally, always practise the 'Safe electrical isolation and lock off' procedure shown in Fig. 3.11*

Assessment questions

Identify the statements as 'true' or 'false'. If only part of the statement is false, tick false.

1   The 'environment' describes the world in which we live
    True ☐   False ☐

2   The WEEE Regulations tell us that we have a 'duty of care' to handle, recover and dispose of all waste responsibly
    True ☐   False ☐

3   The Controlled Waste Regulations tell us to recycle all electrical and electronic equipment
    True ☐   False ☐

4   The Packaging Regulations tell us that the volume and weight of the package shall be a minimum and be either reusable or recyclable
    True ☐   False ☐

5   The Noise at Work Regulations are intended to reduce hearing damage caused by loud noise at work
    True ☐   False ☐

6   The duties of the EHO (Environmental Health Officer) are concerned with eliminating dangers to human health associated with the living and working environment
    True ☐   False ☐

7   A solar thermal hot water heating system turns sunlight directly into electricity
    True ☐   False ☐

8   A solar photovoltaic (PV) system pumps water through a solar panel on the roof to heat the domestic hot water
    True ☐   False ☐

9   A heat pump is used to upgrade low energy heat to a higher temperature
    True ☐   False ☐

10 Biomass is derived from plant materials and animal waste and can be used to generate heat and produce fuel for transportation
    True ☐   False ☐

Multiple choice assessment questions

Tick the correct answer. Note that more than ONE answer may be correct.

11 Our surroundings and the world in which we live is one definition of:

   a.   the Health and Safety at Work Act ☐

   b.   the Building Regulations ☐

   c.   the environment ☐

   d.   the water table ☐

12 Environmental technology systems

   a.   are eco friendly ☐

   b.   use renewable energy ☐

   c.   use fossil fuel ☐

   d.   use nuclear energy ☐

13 Identify the hazardous materials below:

   a.   old glass bottles ☐

   b.   old fluorescent tubes ☐

   c.   used batteries ☐

   d.   off cuts of trunking and conduit ☐

14 Identify the recyclable materials below:

   a.   old glass bottles ☐

   b.   old fluorescent tubes ☐

   c.   used batteries ☐

   d.   off cuts of trunking and conduit ☐

15 The Packaging Regulations tell us that all packaging must be designed and manufactured so that the:

   a.   goods can never be broken ☐

   b.   volume and weight are at a minimum ☐

   c.   packaged goods can be moved with a forklift truck and so avoid manual handling ☐

   d.   used packaging can be recycled and reused ☐

16 Identify four things that you use at work that would require to be disposed of as hazardous waste.

17 Identify six pieces of equipment that would require to be disposed of correctly under the WEEE Regulations.

18 How do the Noise at Work Regulations protect workers?

19 Use bullet points to state the basic operating principles of a solar hot water heating system.

20 Use bullet points to state the applications and limitations of a solar hot water heating system.

21 Use bullet points to state the basic operating principle of a solar photovoltaic system.

22 Use bullet points to state the applications and limitations of a solar photovoltaic system.

23 State the advantages and disadvantages of wind energy generation.

24 Very briefly, in three sentences, describe the basic principle of heat pumps.

25 Very briefly, in three sentences, describe the basic principle of CHP systems.

26 In one sentence describe biomas heating.

27 In one sentence describe micro-hydro generation.

28 What is a 'smart' electricity meter?

29 What is 'rainwater harvesting'?

30 What is 'grey water recycling'?

Solutions to assessment questions

Chapter 1

1.	True	10.	True	19.	c	28.	c
2.	True	11.	a, b, c	20.	c	29.	a
3.	True	12.	d	21.	d	30.	d
4.	False	13.	c	22.	a	31.	d
5.	True	14.	a	23.	c	32.	a
6.	False	15.	b	24.	c	33.	b
7.	True	16.	d	25.	c	34.	c
8.	True	17.	a	26.	a	35.	d
9.	False	18.	b	27.	c		

Chapter 2

1.	True	15.	b	29.	b	43.	d
2.	False	16.	c	30.	a, b	44.	d
3.	True	17.	d	31.	b, c	45.	c
4.	False	18.	c	32.	b	46.	a
5.	False	19.	d	33.	d	47.	b
6.	True	20.	c	34.	b	48.	c
7.	True	21.	c	35.	a	49.	a, d
8.	True	22.	d	36.	a, c	50.	c
9.	False	23.	c	37.	c, d		
10.	False	24.	a, c	38.	c		
11.	c	25.	b, d	39.	b		
12.	b	26.	c	40.	b		
13.	d	27.	b	41.	c		
14.	c	28.	b	42.	d		

Chapter 3

1.	True	16.	c	31.	c	46.	b
2.	False	17.	a, b, d	32.	a, b	47.	b
3.	False	18.	c, d	33.	a, c	48.	a
4.	True	19.	c, d	34.	a,d	49.	d
5.	True	20.	c	35.	d	50.	d
6.	False	21.	d	36.	b	51.	d
7.	True	22.	b, d	37.	c	52.	d
8.	True	23.	b	38.	d	53.	b
9.	True	24.	c	39.	d	54.	c
10.	False	25.	a, d	40.	c	55.	c
11.	b, c, d	26.	d	41.	b	56.	c
12.	c, d	27.	b	42.	a	57.	b
13.	c	28.	a	43.	d	58.	c
14.	a	29.	d	44.	b	59.	c
15.	b	30.	b	45.	d	60.	b

Chapter 4

1.	True	16.	d	31.	c	46.	c
2.	True	17.	d	32.	a	47.	a
3.	True	18.	b	33.	b	48.	b
4.	False	19.	c	34.	b, c, d	49.	d
5.	False	20.	d	35.	c	50.	c
6.	False	21.	c	36.	a	51.	b
7.	True	22.	b	37.	b	52.	b
8.	False	23.	a	38.	d	53.	d
9.	False	24.	a	39.	d	54.	c
10.	True	25.	b	40.	d	55.	b
11.	b	26.	c	41.	b	56.	a, c
12.	a, c, d	27.	d	42.	b, c	57.	d
13.	c	28.	c	43.	a, b, c	58.	c
14.	d	29.	a	44.	b	59.	a, d
15.	b	30.	d	45.	a	60.	b, c

Chapter 5

1.	True	9.	True
2.	False	10.	True
3.	False	11.	C
4.	True	12.	a, b
5.	True	13.	b, c
6.	True	14.	a, d
7.	False	15.	b, d
8.	False	16 to 30	Answers in text

APPENDIX
Environmental
organizations

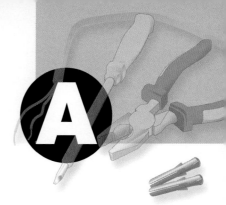

The Department of Energy and Climate Change (DECC) for grants

BS 7671:2008 Engineering Recommendations G 83/1 and G 59/1 Published by the Energy Network Association and the Department for Business, Enterprise and Regulatory Reform (BERR) for technical specifications.

Building Regulations England and Wales. The Department of Communities and Local Government at www.communities.gov.uk and for Scotland, The Scottish Building Standards Agency at www.sbsa. gov.uk

The Labour Governments "Feed in Tariff" was introduced by The Climate Secretary Ed Miliband to encourage green electricity producers by paying a subsidy for every kWh of electricity generated by renewable energy fed back to the National Grid. The Coalition Government which came to power in 2010, are committed to supporting this policy.

For information on the Feed in Tariff Scheme see the Office of the Gas and Electricity Markets (OFGEM) website at www.ofgem.gov.uk/fits

Energy Saving Trust at (020 7222 0101) and www.energysavingtrust. org.uk for advice on grants and products.

Micro-generation product advice and their own Certification Scheme at (01752 823 600) and www.microgeneration.com

The Carbon Trust at www.carbontrust.co.uk/energy offers free advice on loans to businesses who are upgrading to more energy efficient equipment. The size of the loan will depend upon the CO2 savings.

Planning Guide for Solar, PV and wind turbine installations can be found at www.planningportal.gov.uk/uploads/hhg/houseguide.html

Best practice guide for installing micro-generation systems can be found on the Electrical Safety Council website at www.esc.org.uk/ bestpracticeguides.html

Rainwater harvesting guidance and products can be found in abundance by Googleing "rainwater-harvesting".

Index

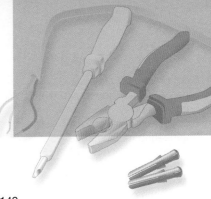